Introducing Genetics

From Mendel to molecule

Introducing Genetics

From Mendel to molecule

Alison Thomas

First published by Nelson Thorne 2003
Reprinted by Taylor & Francis Group 2005

A CIP catalogue record for this book is available from the British Library.

ISBN 0 7487 6440 2

Taylor & Francis Group
4 Park Square, Milton Park, Abingdon, Oxon, OX14 4RN, UK.
World Wide Web home page: www.tandf.co.uk

Distributed in the USA by: Fulfilment Center, Taylor & Francis,
10650 Toebben Drive, Independence, KY 41501, USA.
Toll Free Tel: +1 800 634 7064;
E-mail: taylor and francis@thomsonlearning.com

Distributed in Canada by: Taylor & Francis, 74 Rolark Drive, Scarborough,
Ontario M1R 4G2, Canada.
Toll Free Tel: +1 877 226 2237;
E-mail: tal_fran@istar.ca

Distributed in rest of the world by: Thomson Publishing Services, Cheriton House,
North Way, Andover, Hampshire, SP10 5BE, UK.
Tel: +44 (0)1264 332424;
E-mail: salesorder.tandf@thomsonpublishingservices.co.uk

Illustrations by Donald Stubbs

Typeset by Saxon Graphics Ltd, Derby, UK

Printed and bound by TJ International, Cornwall, UK

Cover photograph: Pasieka/Science Photo Library
DNA molecule (down right), computer artwork. DNA (deoxyribonucleic acid) consists of two strands of sugar phosphates forming a double helix (spiral). The strands are linked by nucleotide base pairs (horizontal struts). The sugar-phosphate strands are blue, but highlighted orange at upper right. In this highlighted section, the nucleotide bases are blue. There are four types of nucleotide base found in DNA. The sequence of these bases along the molecule is the genetic code. This DNA genetic code controls growth and development, and it is passed on during reproduction. DNA is found in all living things, and is stored in cell nuclei.

Contents

Acknowledgements

I would like to acknowledge the tremendous support and encouragement that I have received throughout the development and writing of this book from my editor, Catherine Shaw. Special thanks are also due to Chris Wortley, Production Editor; Sara Hulse, copy editor and Don Stubbs for his enlivening cartoons and illustrations. I am also extremely grateful for the valuable discussions, comments on the manuscript and other help that I have received from colleagues and friends, in particular from Toby Carter, April Erekson, Judy Forshaw, Nancy Harrison, Dawn Hawkins, Guy Norton, Sheila Pankhurst, Helen Roy and Anne Taylor.

The author and publishers are grateful to the following for permission to reproduce copyright material. Every effort has been made to trace all relevant copyright holders, but if any have been inadvertently overlooked, the publishers will be pleased to make the necessary arrangements at the first opportunity.

Biofotos Fig. 10.9; Biophoto Associates Fig. 5.5, 5.10, 6.3, 7.7, 8.4; Dr Toby Carter Fig. 4.14; Fig. 6.10 is Crown Copyright; April Ereckson Fig. 9.3; Judy Forshaw Fig. 2.7; Dawn Hawkins Fig. 9.1; © Houseman, University of Ottawa; Carol Farnei/Planet Earth Pictures Fig. 6.5; Doug Lundberg, United States Air Force Academy, Colorado; Figs 1.3, 5.1; The Master and Fellows of Gonville and Caius College, Cambridge Fig. 2.5; Dr Gopal Murti Fig. 13.3; www.lathyrus.com; The Roslin Institute Fig. 1.5; Mark Rowland Fig. 4.8; Science Photo Library Fig. 5.6, Box 7.1 (right), 11.5, 12.11, 13.5 (© Hank Morgan); St John's College Library, Cambridge Fig. 4.4; Institute for Molecular Virology & Department of Biochemistry, University of Wisconsin-Madison Fig. 13.4.

The following line drawings were previously published in Nelson Thornes books as detailed. The figure in Box 11.1 and Figs 11.2, 11.6, 11.10, 12.2 from Brown, T.A. (1998) *Genetics: a Molecular Approach.*; Figs 11.3, 11.7, 11.9 the Figs in Box 12.1, 12.3, 12.7, 12.8, 12.10 from Roberts, M., Reiss, M. and Monger, G. (1993) *Biology: Principles and Processes*; Fig. 12.9 from Williams, G. (2000) *Advanced Biology for You*; Fig. 13.1 from Adds, J. Larkcom, E. and Miller, R. (2001) Nelson Advanced Science, *Genetics, Evolution and Biodiversity*.

For my mother

Margaret Thomas 1933 – 2002

1 Introduction

Each of us starts life as a single fertilized egg that develops, by division and differentiation, into a mature adult consisting of a hundred trillion cells. Each cell is specialized for a particular function – for example, a muscle cell able to contract or a nerve cell programmed to conduct a nervous impulse. The information that guides this carefully orchestrated development and maintains a fully functioning adult is contained in an estimated 30,000 genes, and is itself stored within each and every cell. The discipline of genetics is concerned with attempting to understand the nature of such information, how it is transmitted from generation to generation, and how it is stored, expressed and regulated.

Genetics, as a scientific discipline, is a twentieth century development, although the recognition of the principle of heredity is clearly much older. The development of agriculture and farming, from around 8000 BC, was only possible because people realized that desirable – and undesirable – traits could be passed to successive generations. The domestication of animals and cultivation of plants was achieved by selecting those genetic variants having the characteristics desired by farmers.

Over the centuries there have been many attempts to explain inheritance and the link between succeeding generations, but it was Gregor Mendel

(1822–84) who originated the scientific theory we accept today (Box 1.1). Between 1856 and 1863 he performed a series of breeding experiments with pea plants in the garden of the monastery near Brno in the Czech Republic where he lived as a monk. He concluded that the patterns of inheritance he observed between succeeding generations were only consistent with a **particulate** explanation for inheritance. His theory involved **hereditary factors**, each controlling a separate trait, which passed unchanged from parent to offspring during reproduction. Mendel published an account of his work in 1866. Unfortunately the intellectual climate of that time was not ready to accept this challenge to the then widely held belief in a blending basis to heredity, and his work went largely unnoticed. It was not until 1900, 16 years after Mendel's death, that the significance of his finding was fully

Box 1.1 The life of Gregor Mendel

Gregor Mendel was of humble origins, a peasant's son from the tiny village of Herzendorf in the foothills of Silesia, now on the Czech/Polish border. Herzendorf was unusual amongst such peasant communities in that it had a schoolmaster, who was quick to recognize the gifts of his young pupil. He persuaded Mendel's parents to send their son to a high school in the nearby town of Troppau. It was a struggle for Mendel's parents to find the fees, but they were rewarded by their son leaving the school in the summer of 1840 with a report marked 'super-excellent' in all subjects – except religion! By now Mendel had his heart set on an academic career. He enrolled at the nearby pre-University and tried to secure work as a private tutor in order to support himself, but it was impossible without friends and recommendations. In despair he turned to one of his teachers, Professor Franz. By happy coincidence, Franz had been asked to recommend candidates for the novicehood at the monastery at Brünn, Moravia (now Brno) and had no hesitation in putting Mendel's name forward. Gregor Mendel was admitted to the monastery as a novice on 9 October 1843.

Recognizing Mendel's academic gifts, the monastery arranged for their novice to teach at the local secondary school. For the next 20 years Mendel was a much respected master of physics and natural history. In his spare time, he began experimenting in the monastery garden. Mendel had become interested in heredity; in the variability shown by members of a given species and how it is passed from generation to generation. Hybridization was a fashionable area of research for mid-nineteenth century botanists and zoologists. Mendel possessed a good working knowledge of current literature and soon realized that the uniformly bewildering results that researchers obtained was because they tried to be all-inclusive, considering the inheritance of all observable traits in every experiment. Mendel, by contrast, believed the only way to obtain a clear answer from Nature was to pose a single simple question, to follow the inheritance of just one clear-cut feature. Thus began his decade of experiments with the garden pea.

Before Mendel had an opportunity for any further work he was appointed abbot of the monastery. Inevitably Mendel now found that much of his time and energies were engaged in public duties, particularly after 1874 when he became embroiled in a fierce struggle with the government over the taxation of monasteries. He was also becoming increasingly troubled by the kidney problems that led to his death in 1884, at the age of 62.

appreciated. Three botanists, Hugo de Vries (1848–1935), Carl Correns (1864–1933) and Erich von Tschermak (1871–1962), each independently rediscovered Mendel's paper when analysing the results of their own similar breeding experiments and immediately recognized the importance of his work. In the intervening years since Mendel's experiments there had been progress in other complementary areas of biology. Improvements in microscopy had revealed the presence of discrete chromosomes in the cell nucleus, and it was quickly realized that the transmission of chromosomes during cell division and reproduction exactly paralleled the behaviour of Mendel's hereditary factors. A particulate basis to heredity was now plausible and the science of genetics was born.

Genetics is a huge and varied discipline, which approaches inheritance from various aspects – molecular and cellular, in the individual and in populations. Before considering different aspects of the hereditary process in detail, a simplified overview of key principles will be given, with the aim of introducing some initial vocabulary and a conceptual framework in which to fit ideas as they are considered in subsequent chapters.

1.1 Key concepts in genetics

The physical site of potential heredity in **eukaryotic** organisms (i.e. animals and plants) is the central **nucleus** of each cell (Fig. 1.1). In **prokaryotic** (i.e. bacterial) cells the genetic material is in a less well defined central area, the **nucleoid** (Fig. 1.2). In both eukaryotes and prokaryotes, the genetic material within the nucleus is **DNA**, an abbreviation for deoxyribose nucleic acid. **RNA** (ribonucleic acid) is another nucleic acid, involved in the transfer of information from the nucleus to the cytoplasm and in its expression. Nucleic acids, along with proteins, carbohydrates and lipids, make up the four major classes of biomolecules found in living organisms.

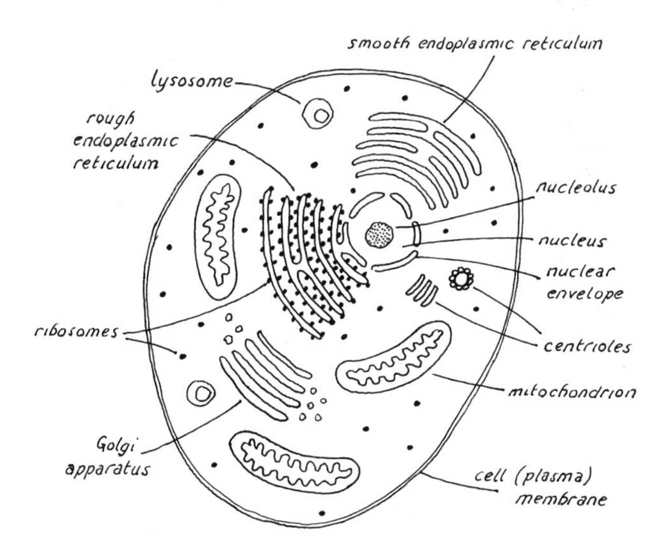

Fig. 1.1
Simplified diagram of an animal cell showing the nucleus and other key organelles.

Fig. 1.2
Simplified diagram of a bacterial cell.

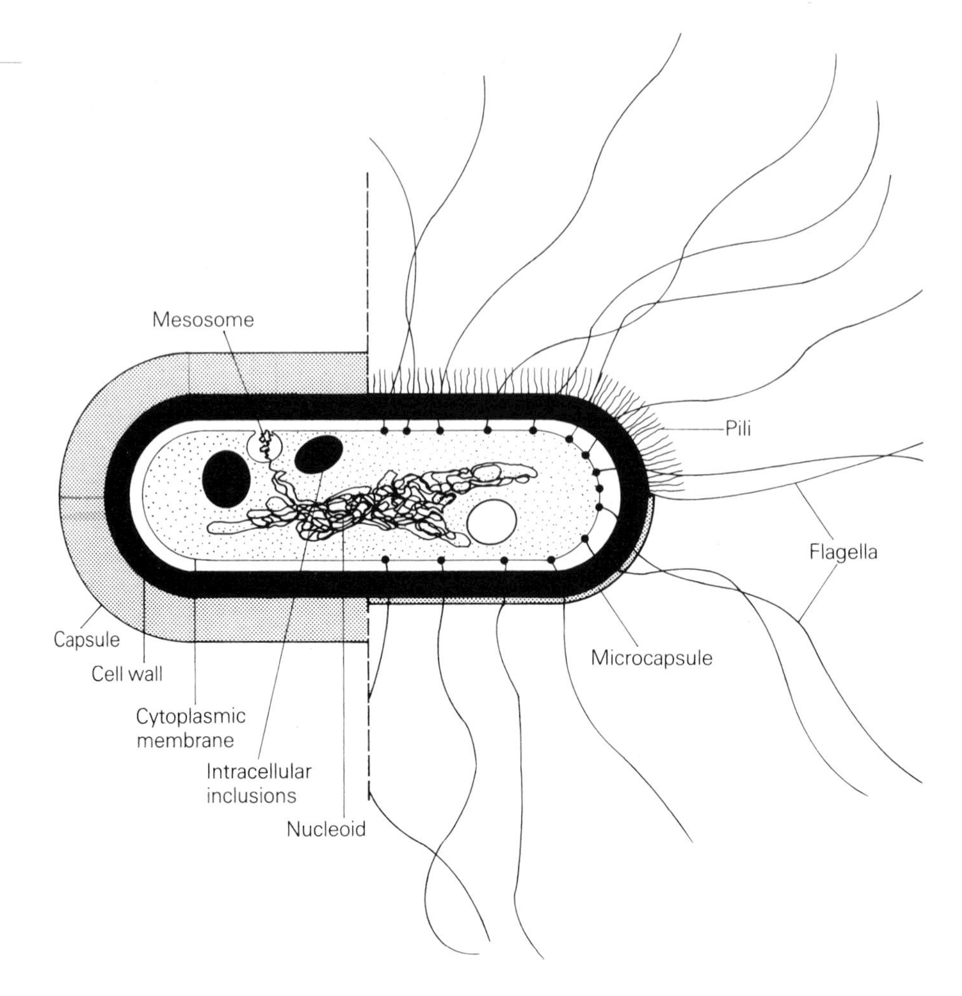

Within the eukaryotic cell nucleus the genetic material is distributed between a number of **chromosomes**. Each chromosome consists of a single linear DNA molecule complexed with protein. By contrast each bacterial cell contains a single, circular strand of DNA. Different regions of a chromosome represent different hereditary units or **genes**. A gene, therefore, can be defined in functional terms as an informational storage unit that can be replicated, expressed and regulated. From a chemical perspective it is simply a piece of DNA.

Chromosomes are only easily visualized when nuclei are dividing as a prelude to cell division, i.e. during the processes of **mitosis** and **meiosis**. Mitosis occurs when the genetic material is exactly duplicated and distributed to two new nuclei during normal cell growth. Meiosis refers to the process that halves the amount of genetic material in a cell – generally when gametes are forming prior to sexual reproduction. At such times of nuclear division, the DNA and associated protein of a chromosome tightly coil and condense, giving rise to the characteristic image of a chromosome (Fig. 1.3), visible under a light as well as an electron microscope. Between cell divisions, chromosomes exist in an uncoiled state, known as **chromatin**, which can only be studied under an electron microscope.

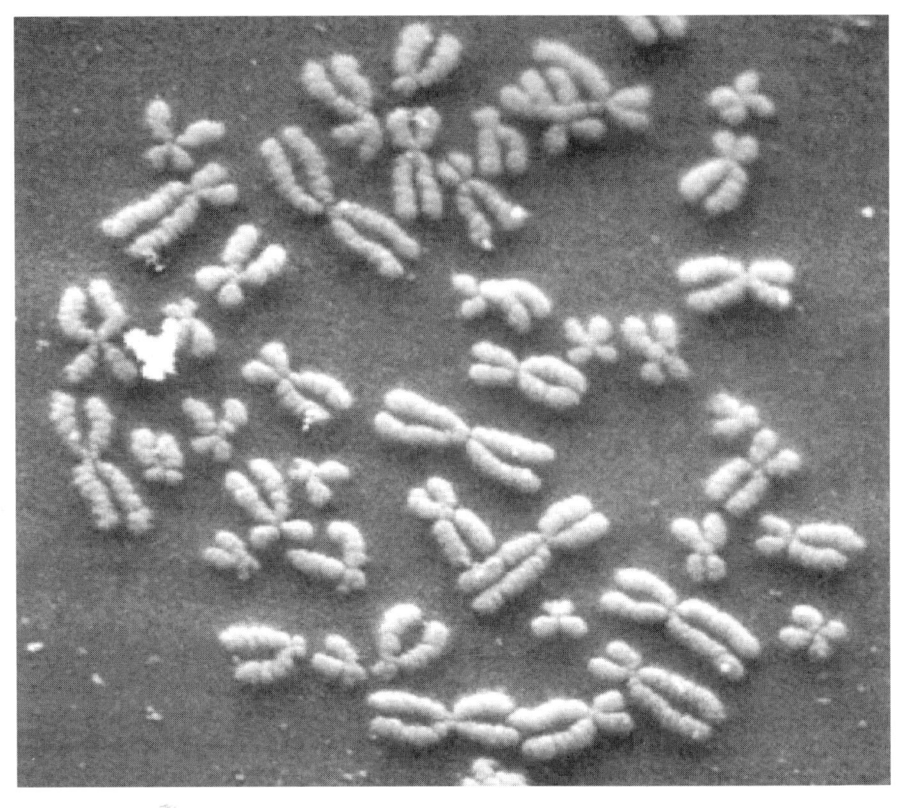

Fig. 1.3
Scanning electron micrograph of human mitotic chromosomes.

The chemical building blocks of DNA are called **nucleotides**, of which there are four types. The key to a gene's storage of hereditary information lies in the precise sequence of nucleotides making up a gene. Expression of a gene's stored information leads directly to the synthesis of a protein. The coded information within DNA is first transferred, during a process called **transcription**, into an RNA molecule. This RNA molecule then associates with a cellular organelle, the **ribosome**, where it directs synthesis of the encoded protein molecule, during a process called **translation** (Fig. 1.4). Proteins perform a range of diverse roles within living organisms. Many are enzymes, catalysing biological reactions. Others perform structural, immunological, nervous or hormonal roles. Because genes encode proteins, one goal of the study of inheritance is to understand how the great diversity of inherited traits can be related to the working of proteins.

The study of genetics, therefore, involves all living organisms and encompasses all levels of biological organization, from molecules to populations. In seeking to understand ways in which genes determine traits, a range of approaches is adopted, producing three main branches of genetics: **transmission**, **molecular** and **population**. The earliest, and 'classical', approach is the study of **transmission genetics**. Insights into genetic principles are gained from studying patterns of inheritance from parents to offspring over several generations. **Molecular genetics** has had the greatest impact on genetic knowledge over the last couple of decades and has led to the development of the discipline of **DNA biotechnology**, where genes controlling specific traits are identified, sequenced, cloned and physically manipulated – all with profound implications for medicine, agriculture and

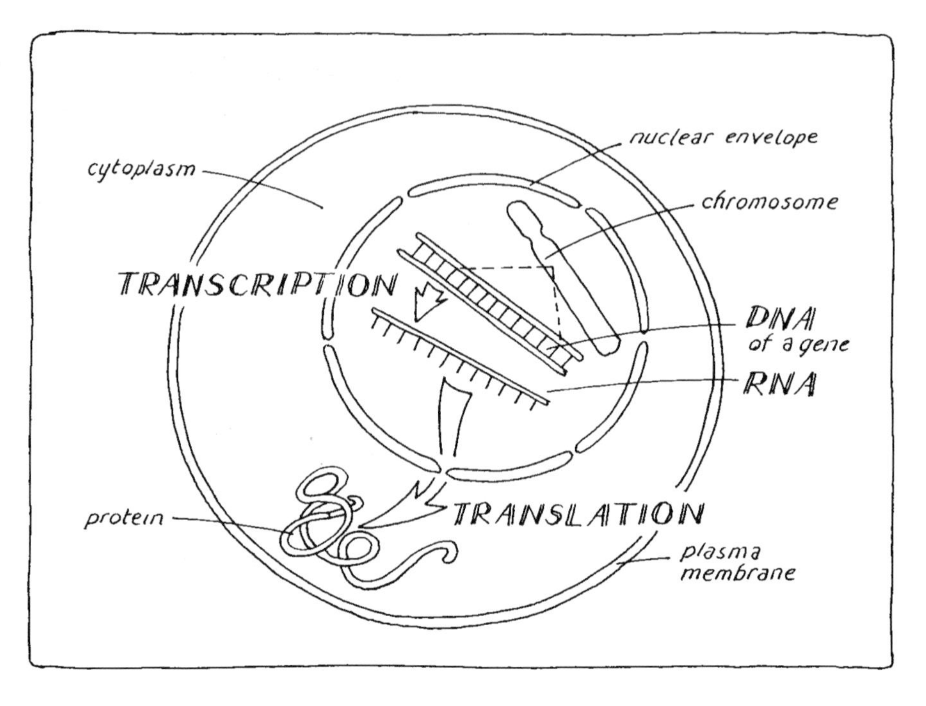

Fig. 1.4
Expression of a gene's stored information: a summary of the steps leading from DNA to proteins.

bioethics (Fig. 1.5). **Population genetics** analyses the variation present within populations and considers mechanisms that promote the maintenance of certain kinds of variation at the expense of others. Such information provides possible insights into evolutionary processes. It also enables us to make predictions about future variation, which has implications, for example, for conservation programmes.

Together these various approaches to the study of genetics have produced one of the most advanced scientific disciplines. The aim of this book is to give a balanced, straightforward and informative introduction to this diverse and ever expanding subject, in our current Age of Genetics.

Fig. 1.5
Triumph or nightmare? Dolly, the first cloned mammal.

2 Monohybrid inheritance

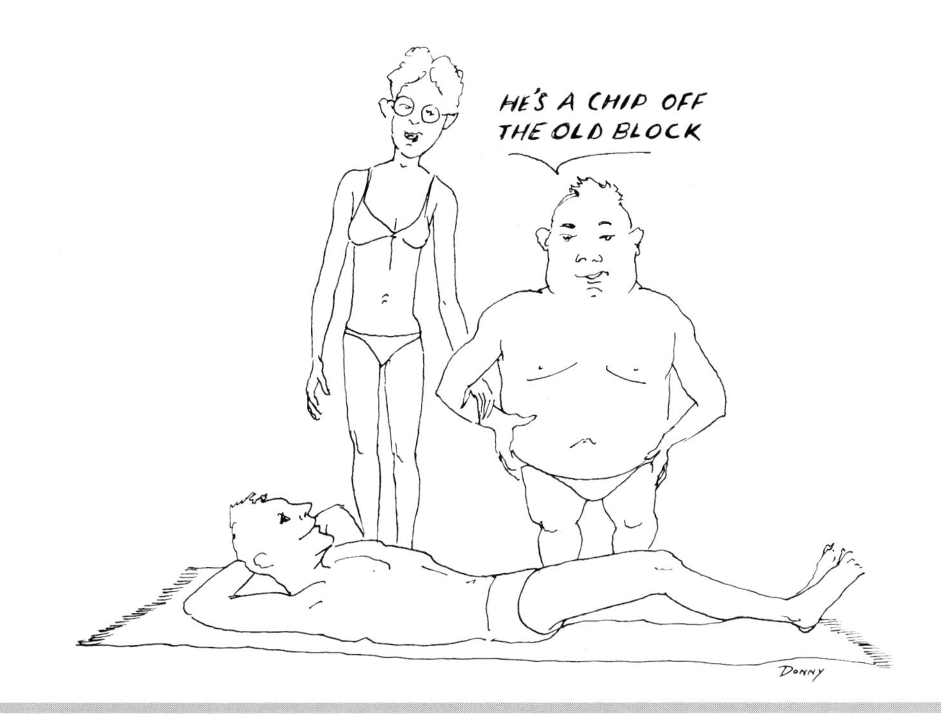

We are all the same – yet different! Look around any crowded high street on a Saturday morning. Everyone is unmistakably human – there are similarities. Yet at another level there is tremendous variability – in height, in skin colour, in face shapes, in build. Some people might have brown hair while others have blonde, some curly, others straight. If you had a sophisticated molecular biology testing kit to hand you would find variations in the activity of particular enzymes, in blood groups and in many other 'hidden' characters. To account for all this variability within a species we can examine the actions of genes. Many of the differences and similarities between individuals can be understood by looking at the transmission patterns of genes from parents to offspring. Certain 'rules of inheritance' exist. Understanding these rules can explain the status quo in any generation and enable us to make predictions about future generations (important in many situations, for example family planning, agriculture, livestock farming, conservation).

Recognising the basic rules of inheritance marked the beginning of the modern discipline of genetics, and is the work for which Gregor Mendel is famed (see Box 1.1). It is important, though, at the start of any study of genetics to point out that an individual's genes cannot totally account for all the details of each structural, biochemical, physiological and behavioural feature that we observe in an organism. Many characteristics are influenced by non-genetic factors, i.e. by aspects of the environment. Body mass is an obvious example. Our attention is constantly being drawn to the drastic effects of overeating and little exercise on human body mass! More controversial are environmental and other influences on behavioural features and personality. It is essential that we recognize that there are other, non-genetic influences acting on organisms and, consequently, that we do not view genetics in a totally deterministic way. Individuality cannot be totally accounted for by an organism's genes. How we may assess the relative contribution of genes and environment to a given trait will be discussed in a later chapter. The aim of this chapter is to introduce:

- some fundamental principles of genetic inheritance;
- the terminology associated with these key principles.

2.1 Key principles of genetic inheritance

These principles explain the expression of a given trait in an individual in one generation and enable predictions to be made about future generations. The best context in which to understand them is that of breeding experiments. One species in which there has been intensive breeding is the tomato (*Lycoperiscum esculentum*). We will consider one characteristic – the colour of its fruit. There are two possible colours – red and yellow. Most commercially grown varieties yield red fruit, but there are also yellow varieties. Two main questions will be addressed in this section:

1. What colour fruits are produced if red- and yellow-fruiting varieties are crossed?
2. How can we explain the consistency of fruit colour from generation to generation in the different varieties?

Firstly we will consider the consequences of crossing a red- and a yellow-fruiting variety. In a breeding experiment of this nature, when the genetic basis of one character is being investigated (a **monohybrid cross**), there is a set procedure:

1. **True** or **pure-breeding** parents are used. This means that each parent comes from a variety in which (a) all members express the same characteristic; and (b) breeding within a variety produces offspring that *all* show the same characteristic as each other and as their parents.
2. **Cross-fertilization** is performed between the pure-breeding parents. This produces the $\mathbf{F_1}$ (or **first filial**) generation. The traits expressed by the F_1 individuals are observed.
3. The F_1 seeds are planted and the resulting individuals crossed among each other to produce the $\mathbf{F_2}$ (or **second filial**) generation. The traits expressed by the F_2 individuals are observed.

4. Large numbers of crosses are performed to produce large numbers of offspring, so that any statistically significant trends can be recognized.
5. **Reciprocal crosses** are performed, i.e. in some crosses the male gamete comes from one parent, while in others from the other parent. This eliminates any gender-related differences.
6. Precautions are taken to ensure that only the desired fertilizations take place. For example with plants, the flowers are wrapped in muslin bags to prevent 'foreign' pollen gaining access to the stigma. Box 2.1 is a reminder of the basic reproductive process in higher plants.

Figure 2.1 shows the results of performing a standard monohybrid cross as outlined above to investigate the genetic basis of fruit colour in tomatoes.

Box 2.1 Reproduction in higher plants

The flowers of most higher plants are hermaphrodite, i.e. they contain both male (the stamen) and female (the carpel) reproductive organs.

Sexual reproduction involves pollen, containing the male gamete, being transferred by wind or insects to a receptive female structure, the stigma. A tube then grows out from the pollen grain and through the female tissues. This provides a pathway for the male gamete to reach the eggs in the ovary. Generally, many pollen grains land on one stigma. Thus many pollen tubes grow towards many eggs. The eventual result is many fertilizations, and so many seeds within one fruit – as in the tomato.

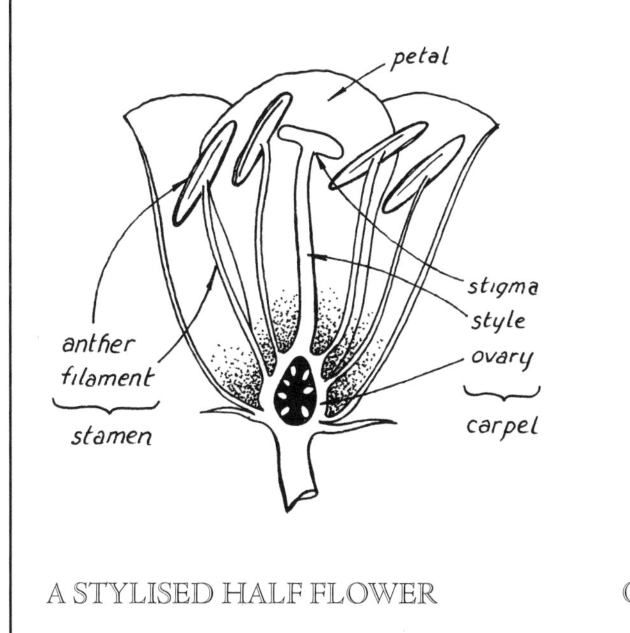

A STYLISED HALF FLOWER

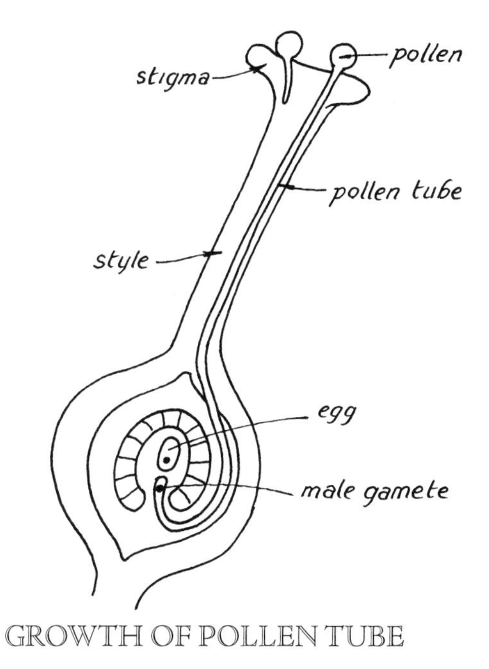

GROWTH OF POLLEN TUBE

Fig. 2.1

Investigating the inheritance of tomato fruit colour.

Parental cross	pure-breeding red fruiter	×	pure-breeding yellow fruiter
F_1 result		**all red fruiters**	
F_2 result		**75% red fruiters, 25% yellow fruiters**	

2.2 *A genetic explanation of the monohybrid cross*

The most striking feature of this monohybrid tomato cross is probably the F_1 result. All F_1 plants produce red fruits. This is true whichever way the cross is carried out – whether the parental pollen comes from the red- or the yellow-fruiting variety. Yet when the F_1 plants are self-fertilized, yellow-fruiting plants are again present in the next generation. These results illustrate some key principles of genetic inheritance.

Within the chromosome set of the tomato there is a gene carrying the instructions for fruit colour. A normal working cell of the tomato contains two copies of the fruit colour gene. One copy will have come via the female gamete and the other via the male. As a prelude to sexual reproduction an organism produces gametes. The cellular process of meiosis (see Chapter 5) ensures that each gamete contains just one copy of each gene so that at fertilization there is once again two copies per trait and normality within cells is maintained. We can now explain the tomato cross.

Genes can exist in different forms or **alleles**. In the current example there are, therefore, two alleles for fruit colour – one is responsible for red fruit and the other for yellow. The original pure-breeding red-fruiting parent had two copies of the red allele in each cell and produced only 'red' gametes. Similarly the yellow-fruiting parent possessed only yellow colour alleles in its cells and in any gametes it produced. Now consider what happens when these pure-breeding varieties are crossed. Red and yellow fruit colour alleles come together in the cells of one individual, but only one colour is expressed. In this example it is red colour that is expressed and yellow that is masked. Such masking is a fundamental genetic phenomenon and is called **dominance**. The expressed allele is the **dominant** one and that silenced the **recessive** allele.

You may have noticed that dominant expression occurs when alleles are the same (both red) or different (one red and one yellow); but recessive expression only occurs in the presence of two identical recessive alleles. This observation introduces two more genetic terms. A **homozygote** refers to an individual that possesses two identical alleles. When the alleles are different an individual is said to be a **heterozygote**. Hence we have the terms **homozygous dominant**, **heterozygous dominant** and **homozygous recessive**. These three terms represent different **genotypes**.

Genotype and **phenotype** are two other commonly used genetic terms. Genotype refers to the particular gene(s) associated with a given trait and phenotype to the physical characteristic that results from expression of the gene(s). Thus, individuals with a homozygous recessive genotype produce the yellow-fruiting phenotype. Genotype, therefore, determines phenotype. Sometimes these terms are used in a broader sense, where phenotype refers to the sum of all characters that an individual possesses, i.e. structurally, biochemically, physiologically and behaviourally; and genotype alludes to the full complement of an individual's genes.

The preceding paragraphs explain the results of the tomato cross and introduce some genetic terminology, summarized in Fig. 2.2. Section 2.3 presents a shorthand way of representing crosses.

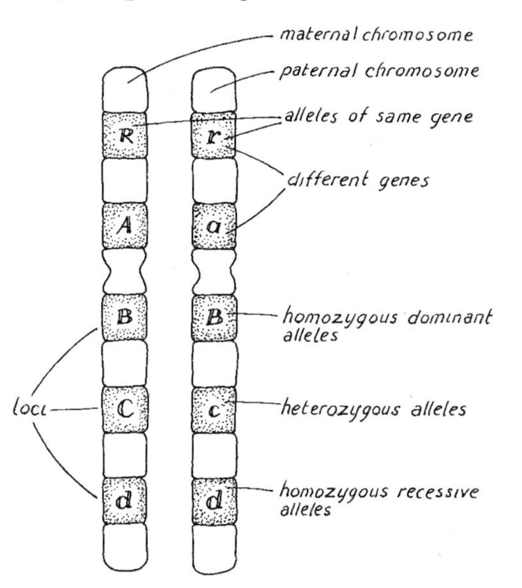

Fig. 2.2
A summary of the relationship between genes and chromosomes. The loci (singular locus) are the positions along the length of a chromosome where particular genes are found.

2.3 Representing genetic crosses

Genes are represented by letters of the alphabet: upper case for the dominant allele and lower case for the recessive allele. The letter chosen is generally the first letter of the dominant phenotype. Thus, **R** can be used to represent the allele for red fruit and **r** the allele for yellow fruit and the production of the F_1 and F_2 generations represented as shown in Figs 2.3 and 2.4.

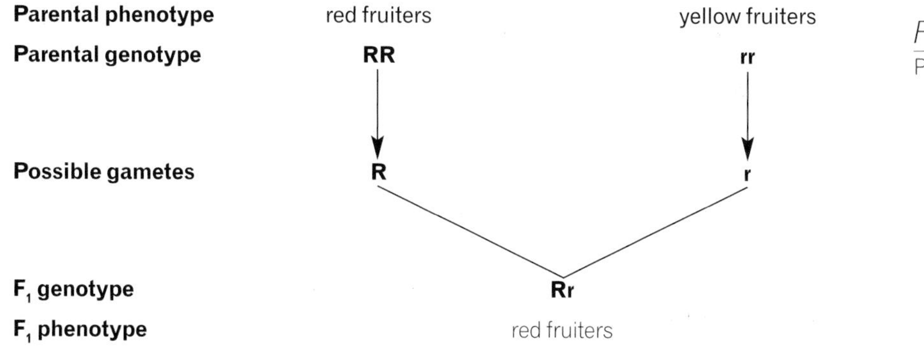

Fig. 2.3
Producing the F_1 generation.

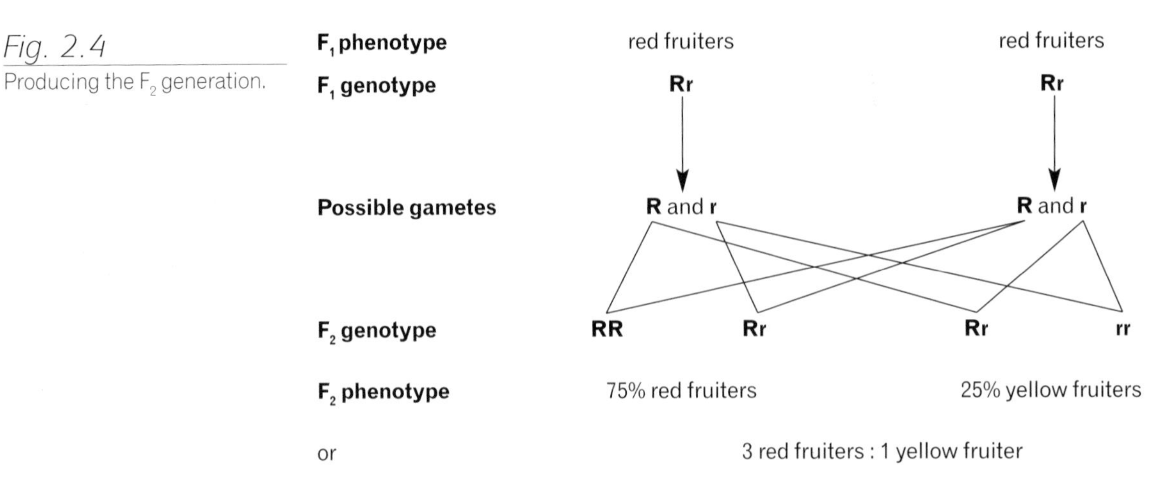

Fig. 2.4
Producing the F_2 generation.

When representing the results of genetic crosses we tend to express relative numbers of different phenotypic classes as **ratios**. Offspring ratios often give valuable clues to an individual's genotype and to which of two alternative characters is dominant.

A few comments are needed about representing fertilizations where an individual produces more than one type of gamete, as in the F_1 red-fruiting plants. The lines linking gametes from plant 1 with those from plant 2 show all the possible fertilizations and assume an equal chance or probability of all events. Consider plant 1 – an **R**-bearing gamete can combine with an **R**- or **r**-bearing gamete from plant 2. Over a large number of fertilizations we could expect 50% of each type of fertilization, i.e. 50% homozygous red plants (**RR**) and 50% heterozygotes (**Rr**). Likewise, the **r**-bearing gamete has an equal chance of combining with an **R**- or **r**-bearing gamete from plant 2, so producing 50% heterozygotes (**Rr**) and 50% yellow homozygotes (**rr**). This produces a theoretical phenotypic ratio of three red-fruiting plants to every yellow-fruiting plant. Rarely, though, do the different types of fertilization occur in exact theoretical proportions. So when we analyse results, we are looking for phenotypic ratios that approximate to 3 : 1 or some other informative relationship.

2.4 A Punnett Square

Instead of using lines to represent possible fertilizations, many geneticists prefer to use a grid or **Punnett Square**, named after the geneticist Reginald Punnett (1875–1967) who first used this method (Fig. 2.5). By using a Punnett Square (Fig. 2.6), there is probably less chance of making a mistake (e.g. of forgetting a possible fertilization) than by drawing lines between gametes, as in Fig. 2.4.

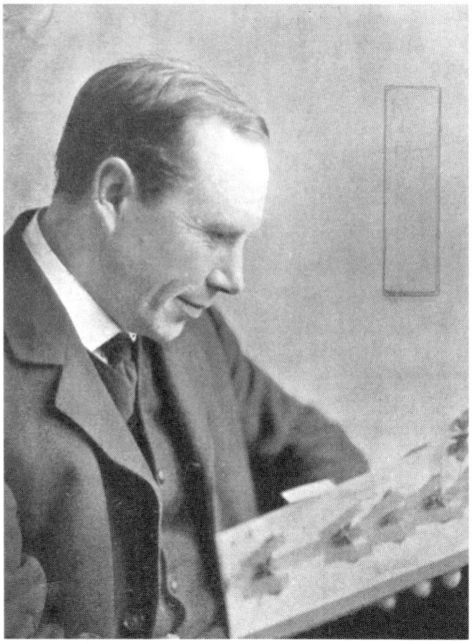

Fig. 2.5

Reginald Punnett. During the early 1900s Mendel's work found a receptive audience among British biologists, including Punnett. His work with sweet peas, poultry and other animals earned him the country's first professorship in genetics, at Cambridge in 1912. Keen to see a practical application of genetics he developed many new breeds of sweet peas, maize and poultry. His *Heredity in Poultry* was the standard work on poultry genetics for several decades.

F_1 plant 1 — **Rr** → **Possible gametes**: **R** and **r**

F_1 plant 2 — **Rr** → **Possible gametes**: **R** and **r**

F_2 result

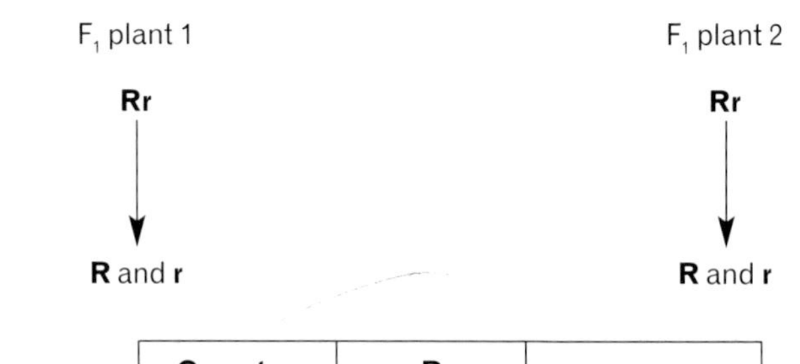

Gametes	**R**	**r**
R	**RR** red fruiter	**Rr** red fruiter
r	**Rr** red fruiter	**rr** yellow fruiter

3 red fruiters : 1 yellow fruiter

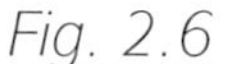

Fig. 2.6

Using a Punnett Square to predict the results of interbreeding F_1 plants.

2.5 Mendel's first law of segregation

The results of crossing pure-breeding red and pure-breeding yellow-fruiting tomatoes closely mimics experiments performed by the 'Father of Genetics', Gregor Mendel. Mendel's contribution to genetics is so important that the adjective 'Mendelian' is often used to describe the kind of experiments that he originally carried out and the principles he formulated.

In his series of carefully planned experiments in his monastery garden at Brno, Gregor Mendel laid the foundations of the modern science of genetics. He worked with the garden pea (*Pisum sativum*) and followed the inheritance of various pairs of contrasting characters (for example round or wrinkled seeds, white or purple flowers, green or yellow seeds and long or short stems) through many generations. He carefully counted the numbers of individuals in each generation showing different alternative phenotypes. The validity of his findings is strengthened by the large numbers of plants with which he worked (often thousands). Although not expressed in the genetic language that we use today, he drew the same conclusions (summarized in Table 2.1) from his experiments as discussed previously with respect to inheritance of fruit colour in tomatoes. Mendel's conclusions led him to propose the first of two important laws, **The Principle of Segregation**, which states that:

> *In the formation of gametes the paired hereditary determinates separate (segregate) in such a way that each gamete is equally likely to contain either member of a pair.*

Table 2.1

A summary of Mendel's conclusions

Traits of an organism are determined by particulate factors
Each parent has two of these particles
Pure-breeding strains contain a pair of identical particles
F_1 hybrids have two different particles
Only one of a pair of different particles are expressed in the F_1 hybrid
Particles are transmitted from parents to progeny through gametes
Each gamete contains just one particle
It is random which one of a pair of particles enters a gamete
Gametes, and so particles, randomly unite in a zygote

Mendel's 'hereditary determinants' are what we now call alleles and, as discussed in Section 5.10, meiosis ensures that each gamete contains just one of a pair of alleles. Fertilization restores the normal paired situation.

2.6 Predicting the outcome of crosses

By considering the pattern of inheritance of fruit colour in tomatoes, certain fundamental rules of inheritance have been illustrated. A knowledge of these rules (that, for example, a dominance relationship exists between alleles, that crosses involving two heterozygotes produce offspring in a 3 : 1 phenotypic ratio) enables us to make predictions about likely phenotypes and genotypes in future generations. This can be very helpful – for example for prospective parents to know the likelihood of any children expressing a given disease.

It is important, however, to remember that genetic predictions are not absolute; they give us only the **chance** or **probability** of certain outcomes. Parents can know, for example, that they each have a genotype that means they have a 1 in 4 chance of having a child with cystic fibrosis; or a breeder of guinea pigs might expect half the offspring of a given mating to be long-haired; but neither of these two outcomes are certainties. It is essential to appreciate this probability or chance aspect to inheritance. A major part of transmission genetics involves making predictions about the chances of certain outcomes over others. Returning to the example this chapter has been using, when the F_1 red-fruiting tomatoes were crossed among each other, the outcome was presented as ¾ red-fruiting and ¼ yellow-fruiting plants in the F_2 generation. This was only a prediction in an ideal world. The observed numbers are more likely to only approximate to a 3 : 1 relationship.

Before introducing any further genetic principles it is necessary to stop and consider what is meant by 'chance' and 'probability'; why, practically, you could plant, say, 20 seeds from an F_1 cross and not get a single yellow-fruiting tomato plant, although theory predicts you would get five.

2.7 Chance and probability in genetics

When introducing ideas about probability it helps to use familiar situations; for example, when tossing a coin we all know that it is equally likely that heads or tails will be uppermost. The probability of throwing a head is thus described as 1 out of 2 or ½. This is defined formally as:

$$\textit{Probability of a particular outcome} = \frac{\textit{number of ways of getting a particular outcome}}{\textit{total number of possible outcomes}}$$

So, with our coin example:

- there is only *one* way of obtaining heads
- there are *two* possible outcomes (heads and tails)
- therefore, the probability of obtaining heads = ½

Consider another, often cited example – the pack of playing cards. What would be the probability of obtaining a seven? The answer is 1/13, by the following reasoning:

- there are *4* ways of obtaining seven (4 sevens in a pack)
- there are *52* possible outcomes (52 cards in a pack)
- therefore, the probability of obtaining a seven = 4/52 = 1/13

Referring back to our F_1 genetic cross (Fig. 2.4), we could ask what is the probability of obtaining a yellow fruiter:

- there is *one* way of obtaining a yellow fruiter (**r** × **r**)
- there are *four* possible outcomes (**R** × **R**; **R** × **r**; **r** × **R**; **r** × **r**)
- therefore, the probability of obtaining a yellow fruiter = 1/4

One other idea should be appreciated here. When considering the number of possible outcomes, each is **equally likely**. You are equally likely to throw a head or a tail, or to pull any one of the 52 different cards from the pack. Each heterozygous F_1 red fruiter produces equal numbers of **R** and **r** gametes, and each is equally likely to be used at fertilization.

2.8 The importance of large numbers

Probability gives us theoretical expectations on the premise that all possible outcomes are equally likely. Consider the case of two parents who already have five daughters and desperately want a son (Fig. 2.7). Surely the next conception will give them their longed for son. After all with six children they could have expected three daughters and three sons. But no – the sixth child is another daughter! Reality often deviates from expectation, and so it

Fig. 2.7
A family of five daughters.

is with genetic experiments. Consider again the F_1 red-fruiting heterozygote. There is no guarantee that it will produce exactly equal numbers of **R** and **r** gametes; there might be, for example, differences in gamete viability. Furthermore, random fertilization rarely uses each kind of gamete exactly equally.

To illustrate this latter idea, consider again the act of throwing a coin. The expected outcome is an equal number of heads or tails uppermost – a 1 : 1 ratio. Ten throws of the coin might produce a marked deviation from a 1 : 1 ratio. Yet after 1000 throws the ratio is likely to be very close to 1 : 1. Deviations from the expected are fewer the more times a probability event is performed: the **observed** result is more likely to equal the **expected**. These considerations stress the importance of using large numbers of individuals in genetic crosses, so that offspring ratios are meaningful and valid interpretations can be made. For example, if a cross yields a 3 : 1 ratio of two different phenotypes, we can accept this as a true reflection of the underlying genetics – in this case, that each parent was a heterozygote.

In conclusion:

- Offspring ratios are often enormously important in indicating the genetic basis of inheritance for a given trait.
- To ensure observed ratios are meaningful, large numbers of offspring must be produced (Mendel appreciated this; see Table 2.2).
- We can then match observed ratios to one of a set of expected ratios for different genetic situations.

Parental traits	Number of F_2 progeny	F_2 ratio
Round × wrinkled seeds	5474 round, 1850 wrinkled	2.96 : 1
Yellow × green seeds	6022 yellow, 2001 green	3.01 : 1
Inflated × constricted pods	882 inflated, 299 constricted	2.95 : 1
Green × yellow pods	428 green, 152 yellow	2.82 : 1
Purple × white flowers	705 purple, 224 white	3.15 : 1
Axial × terminal flowers	651 axial, 207 terminal	3.14 : 1
Long × short stems	787 long, 277 short	2.84 : 1

Table 2.2
Results of Mendel's monohybrid experiments (indicating a clear 3 : 1 pattern!)

The 3 : 1 phenotypic ratio is one of a number of useful genetic outcomes. This chapter has focused on this one ratio, important in the context of monohybrid crosses. The 1 : 1 ratio is another very useful monohybrid ratio, discussed below.

2.9 The test cross

Because yellow fruit colour in the tomato is recessive to red fruit colour, we know immediately the genotype of any yellow-fruiting plant – homozygous recessive, **rr**. However, a red-fruiting plant could be homozygous or heterozygous dominant, **Rr** or **rr**. It is possible to distinguish between these alternatives for any red-fruiting plant by crossing the red fruiter with a yellow-fruiting plant. The phenotypes of the offspring indicate the genotype of the red parent. If the **test cross** yields approximately equal numbers of red- and yellow-fruiting plants (indeed any significant numbers of yellow fruiters) this indicates that the red-fruiting plant is heterozygote. Otherwise the conclusion is that the plant is homozygous (Figs 2.8 and 2.9). Thus, crossing an individual of dominant phenotype, but unknown genotype, with a recessive homozygote is a common and useful test of genotype. If a reasonable number of offspring, usually about 50%, show the recessive phenotype, then it can be concluded that the individual showing a dominant phenotype has a heterozygous genotype.

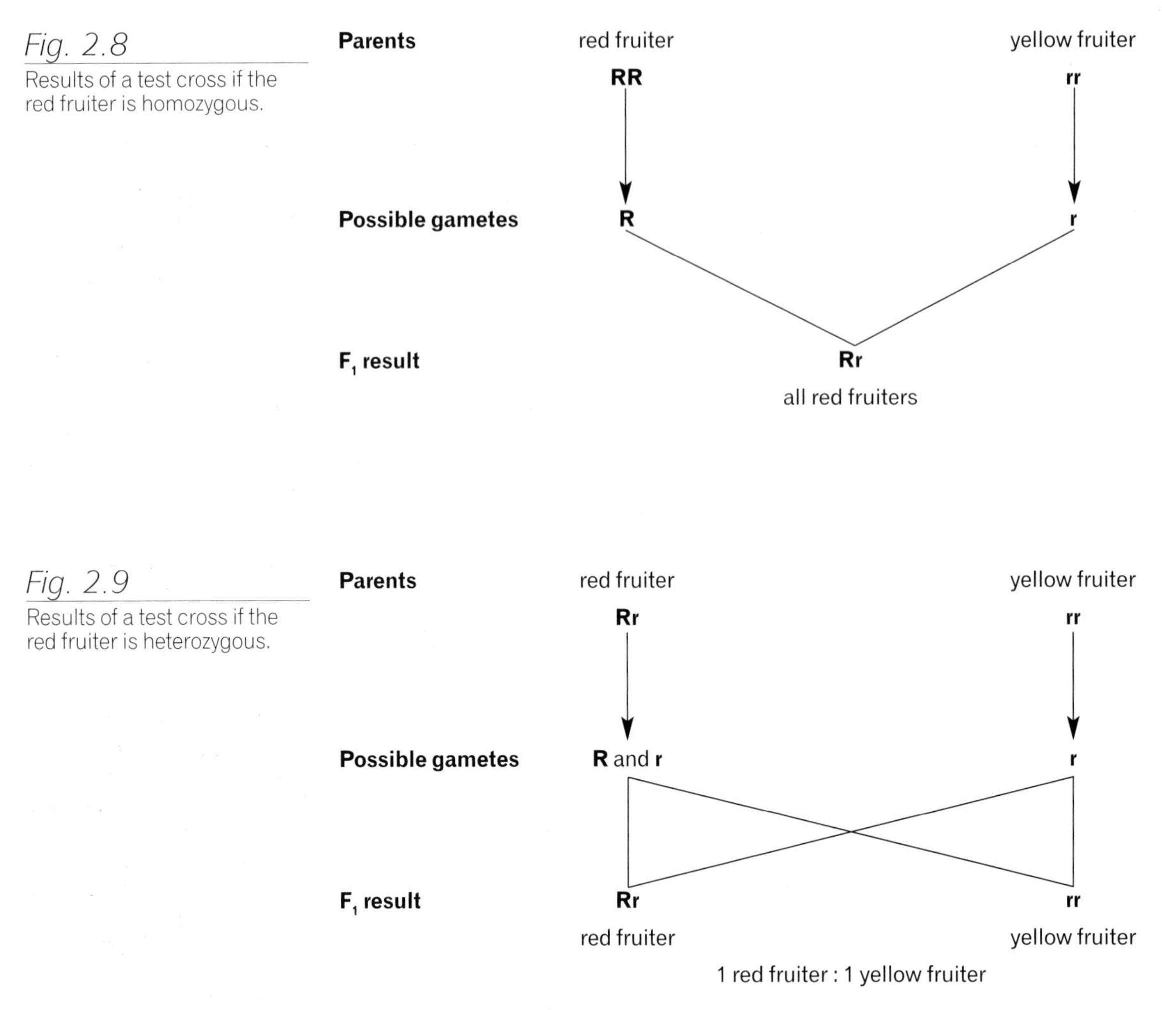

Fig. 2.8
Results of a test cross if the red fruiter is homozygous.

Fig. 2.9
Results of a test cross if the red fruiter is heterozygous.

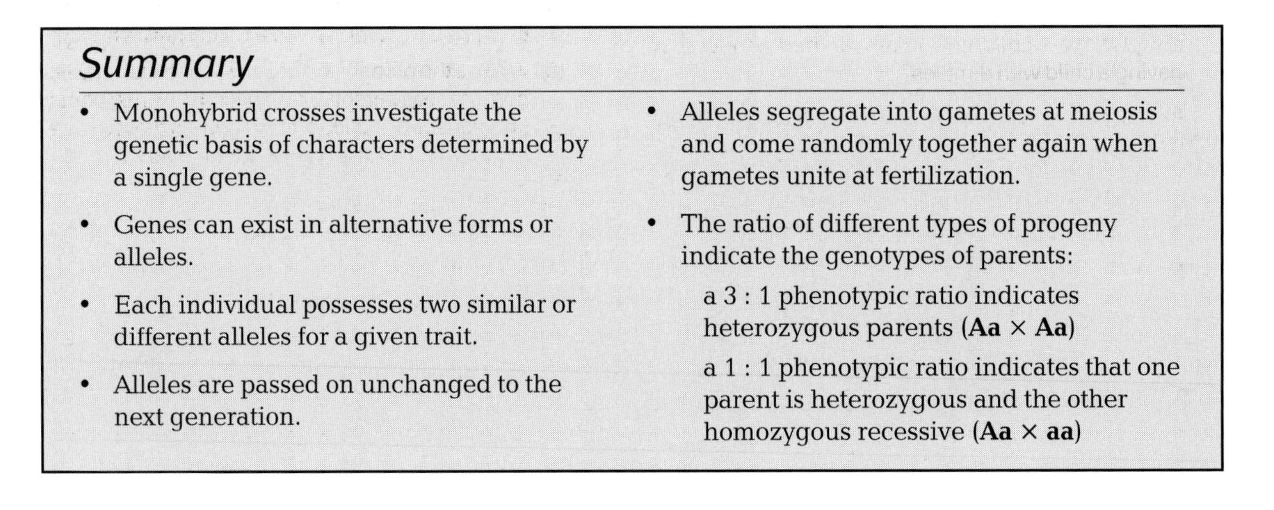

Summary

- Monohybrid crosses investigate the genetic basis of characters determined by a single gene.
- Genes can exist in alternative forms or alleles.
- Each individual possesses two similar or different alleles for a given trait.
- Alleles are passed on unchanged to the next generation.
- Alleles segregate into gametes at meiosis and come randomly together again when gametes unite at fertilization.
- The ratio of different types of progeny indicate the genotypes of parents:

  a 3 : 1 phenotypic ratio indicates heterozygous parents (**Aa** × **Aa**)

  a 1 : 1 phenotypic ratio indicates that one parent is heterozygous and the other homozygous recessive (**Aa** × **aa**)

Problems

1. In mice, black eye colour is dominant to red. A pure-breeding black-eyed male was crossed with a female who was pure-breeding for red eyes.
   (a) What phenotype could you expect in the F_1 generation?
   (b) If the eight F_1 mice were allowed to interbreed, what phenotypes would you expect among the F_2 generation and in what proportions?
2. If an F_2 mouse from Question 1 has black eyes, how can you decide whether it is homozygous or heterozygous?
3. Two short-haired guinea pigs were mated on three separate occasions and produced 21 short-haired and 6 long-haired offspring.
   (a) What is the genetic basis of hair length in guinea pigs?
   (b) What are the genotypes of the short-haired parents?
4. The author Ernest Hemingway was famous for his love of cats: as many as 60 lived with him at his house at Key West, Florida, USA. His cats were also famous – for having six toes. If a normal five-toed stray appeared and mated with one of these six-toed cats, the result was a litter consisting largely or entirely of six-toed cats! What does this suggest to you about the genetic basis of polydactyly in cats?
5. A variety of poppies that are pure-breeding for spots at the base of their petals were crossed with another, non-spotted variety. The F_1 plants were allowed to self-fertilize, with a resulting F_2 generation consisting of 264 spotted and 84 unspotted poppies. What was the phenotype of the F_1 poppies?
6. Dimples are a dominant trait in humans. A man who is homozygous for dimples and a woman without dimples have children. What are their chances of having a child with dimples?
7. In the fruit fly, *Drosophila melanogaster*, grey body colour (**G**) is dominant to black body colour (**g**). A geneticist had three flies with grey bodies, designated P, Q and R. He crossed P and Q and obtained 109 grey-bodied flies. Q and R gave 80 grey-bodied and 28 black-bodied flies; whilst P and R gave 76 grey-bodied flies. What would be the expected genotypic and phenotypic ratios when flies P, Q and R are crossed with black flies?
8. A tobacco grower crossed pure-breeding plants with large leaves with pure-breeding plants with small leaves. The F_1 plants all had large leaves. He allowed these F_1 plants to self-fertilize. The resulting seed, when planted, produced 640 plants. How many of these would you expect to be large-leaved and how many small-leaved?
9. In humans brown eyes is dominant to blue eyes and identical twins occur approximately once in every 300 births. What is the probability of a blue-eyed couple having brown-eyed identical twin boys as their first children?
10. Pecan shells are sometimes thin and liable to shatter during harvest. A farmer wanted to make sure that he wouldn't lose any of his valuable crop through this undesirable trait. Thus, he investigated the genetic basis of this trait by taking pollen from two different trees (A and B), both of which produced thick-shelled pecans, and using it to fertilize flowers on a tree that produced thin-shelled pecans. Pollen from tree A produced 56 plants, of which 29 eventually produced thick-shelled nuts, while pollen from tree B resulted in plants that all produced thick-shelled pecans. What is the mechanism of inheritance of thick and thin pecan shells?
11. In cattle, to be polled (hornless) is dominant to horned.
    (a) What are the genotypes of polled parents that produce a calf which subsequently grows horns?
    (b) What is the probability that any subsequent calves born to these parents are (i) polled; (ii) grow horns; (iii) produce all polled offspring when they mature, regardless of the genotype of their mates?

Extensions to monohybrid inheritance

The analysis of phenotypic ratios among the progeny of a mating provides important clues to the genetic basis of different traits. Chapter 2 introduced this important approach with respect to monohybrid crosses, i.e. to recognizing when a character was determined by one gene with two alleles.

This chapter introduces further ideas about monohybrid inheritance, in particular:

- situations where there are more than two alleles possible at a given locus;
- some differing dominance relationships between alleles;
- how the rules are applied to human inheritance.

3.1 Multiple allelism

As anyone knows who has kept rabbits as pets, they can have a wide variety of different coat colours. There is, for example, the diffuse grey/brown colour of wild rabbits – agouti. This coloration is the result of each hair having a mixture of black and yellow bands. Albino rabbits occur when the hairs lack pigmentation. If a pure-breeding agouti and pure-breeding albino rabbit are crossed the F_1 progeny are all agouti and the F_2 generation shows a ratio of 3 agouti to 1 albino rabbit. Clearly the agouti phenotype is dominant and the two colours are produced by the action of one gene with two alleles: a 'classic' monohybrid situation. That the agouti coloration is dominant to albino may be expected as the variously coloured hairs of agouti rabbits obviously produce a better camouflaged rabbit than a brilliant white one.

Albino rabbits are, however, popular as pets, as are white rabbits with black feet, ears, nose and tail (Himalayans) and the delicate silver-grey chinchillas. If one gene is determining coat colour, how can the normal monohybrid inheritance rules account for these additional variants? This range of rabbit fur colours introduces a new genetic idea – that a gene may have more than two alleles. It can show **multiple allelism**. Although each individual can have no more than two alleles of one gene in its cells, many more can be present among the different members of a population. In fact the main gene determining fur colour in rabbits has four alleles – agouti, chinchilla, Himalayan and albino, with dominance in this descending order.

In other words, agouti is dominant to all other three colours and albino is recessive to them all. Chinchilla, however, is dominant to Himalayan. With these allele relationships four phenotypes can be produced by ten different genotypes, as shown in Table 3.1.

Table 3.1

The genetic control of rabbit coat colour (allele symbols: C = agouti, C^{ch} = chinchilla, C^h = Himalayan, c = albino)

Genotype of rabbit	Phenotype of rabbit
CC	Agouti
CC^{ch}	Agouti
CC^h	Agouti
Cc	Agouti
$C^{ch}C^{ch}$	Chinchilla
$C^{ch}C^h$	Chinchilla
C^{ch}c	Chinchilla
C^hC^h	Himalayan
C^hc	Himalayan
cc	Albino

If the dominant/recessive relationships between the different alleles in a multiple allele series are known, then any cross can be considered in the 'normal' Mendelian way, as developed in Chapter 2. We can, for example, explain to a rabbit breeder the appearance of albino and Himalayan rabbits from a mating between a chinchilla doe and a Himalayan buck when the breeder had been expecting all their rabbits to be chinchillas! Unbeknown to the breeder both the doe and the buck must have been heterozygotes,

with a pure-breeding white-flowering plant. One colour would be expected to be dominant; presumably the stronger one, red. Thus it would be reasonable to assume that all the F_1 plants would produce red flowers. But no! The flowers of the F_1 generation are all pink – neither colour dominates. Instead the red and white traits have apparently blended to give an F_1 generation of intermediate colour to either parent. What happens therefore when these pink F_1 plants are crossed among themselves? The answer is that red-, pink- and white-flowering plants are all observed in the F_2 generation, and in distinctive proportions: one-quarter are red, one-quarter are white and a half are pink-flowering (Fig. 3.4). So, how can we explain these observations?

Fig. 3.3

Red- and white-flowered snapdragons. Various genetic modifications produce a range of other colours.

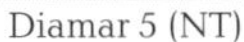

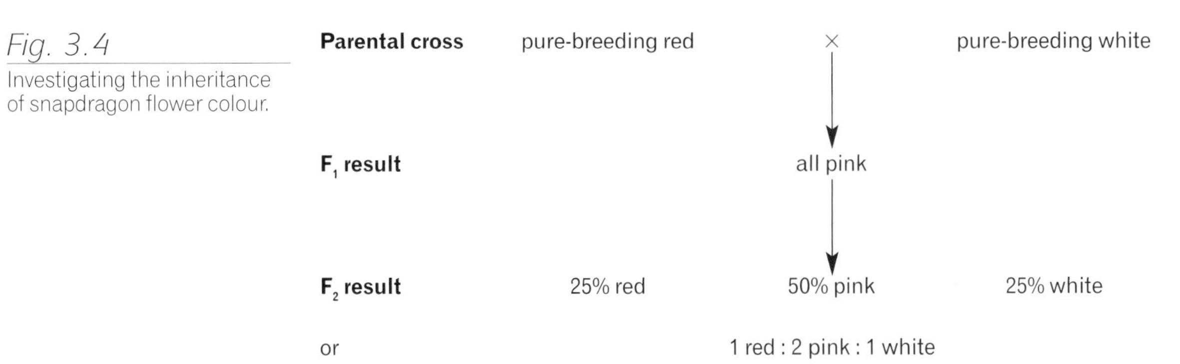

Fig. 3.4

Investigating the inheritance of snapdragon flower colour.

The fact that red and white flowers appear again in the F_2 generation shows that the blending is at the phenotypic level. The red and white alleles have remained as discrete functional units when together in the F_1 plants so that they can segregate at gamete formation and recombine at fertilization to give all possible coloured plants in the F_2 generation. Another genetic phenomenon is illustrated here, that of **incomplete dominance**. This occurs when neither of a pair of alleles shows dominant expression to the other. Thus the heterozygote displays a distinctive phenotype to either of the two homozygotes. Indeed, the heterozygous phenotype is often intermediate in its expression to that of both homozygotes, as in this example of snapdragon flower colour; pink is intermediate in intensity between red and white.

3.3 *Representing crosses involving incomplete dominance*

When displaying genetic crosses it is conventional to use an upper case letter for a dominant allele and a lower case one for the corresponding

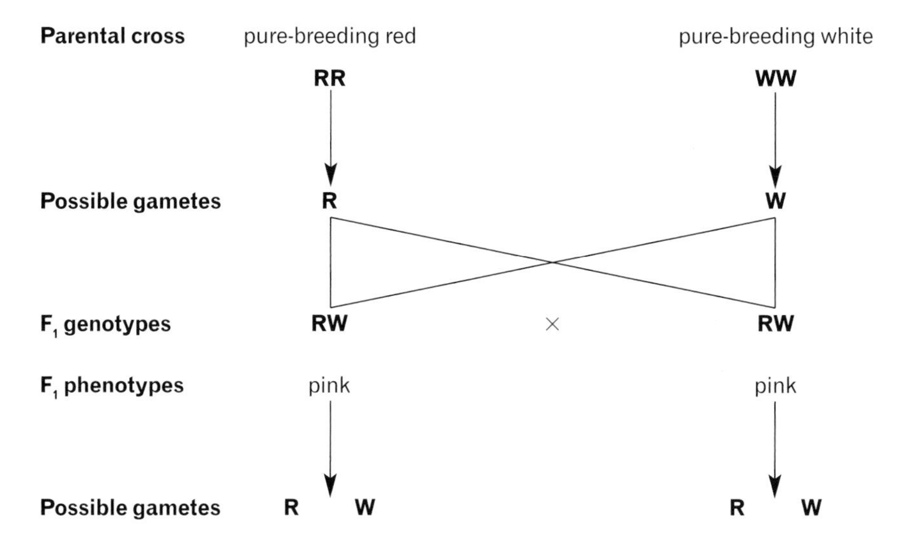

Fig. 3.5
Representing the inheritance of snapdragon flower colour.

F_2 result

Gametes	**R**	**W**
R	**RR**	**RW**
W	**RW**	**WW**

F_2 genotypic ratios 1 **RR** : 2 **RW** : 1 **WW**

F_2 phenotypic ratio 1 red : 2 pink : 1 white

recessive one (Section 2.3). When there is no dominance relationship between two alleles they must be represented differently. Generally a different capital letter is used for each allele. So in the present example of red- and white-flowering snapdragons, **R** might well represent the red allele and **W** the white one. The monohybrid cross could then be represented as shown in Fig. 3.5.

Another informative Mendelian ratio is illustrated. If a cross produces three phenotypes in a 1 : 2 : 1 ratio, then a monohybrid situation involving two incompletely dominant alleles should be suspected, especially if the phenotype of the larger class is intermediate in expression between the other two phenotypes.

3.4 Explaining incomplete dominance

An explanation for incomplete dominance can often be found by considering the nature of the gene product. In such cases only one of the two alleles produces a functional product. This means that the cells of one homozygote produce two units of working product, those of the heterozygote produce one unit and those of the other homozygote none. Consider this interpretation within the context of the various petal colours of snapdragons. Red pigment is formed by a complex sequence of enzyme-controlled reactions. The **R** allele codes for a crucial enzyme in this pathway while the **W** allele fails to produce an active enzyme. Thus, the **RR** homozygote produces sufficient enzyme for a lot of red pigment to be produced from a white precursor. In heterozygotes, **RW**, only one allele is producing active enzyme, so only some of the white precursor pigment is changed. In the petals of these plants both the unchanged white precursor and red pigment is present. As a result the petals appear pink. Any plant of **WW** genotype is, of course, white. This situation is summarized in Table 3.2.

Table 3.2
The relationship between genotype, enzymes and pigment in the determination of snapdragon petal colour

Genotype	State of enzyme	Pigment present	Phenotype observed
RR	100% active	Red	Red
RW	50% active; 50% inactive	Red and white	Pink
WW	100% inactive	White	White

3.5 Redefining dominance relationships

It is common for the heterozygote to show an intermediate phenotype to the two homozygotes when colour phenotypes are being considered. Various other examples are given in the problems at the end of the chapter. Frequently the different colours can be related to varying levels of functional product produced by two alleles, as in the snapdragon example. Indeed, in recent years, as analysis of an increasing number of phenotypes at the biochemical level has become possible, a redefinition of dominance relationships between pairs of alleles has been necessary.

Complete dominance of allele **A** over allele **a** means that the genotypes **AA** and **Aa** are indistinguishable phenotypically. However it is becoming clear that the ability to distinguish the homozygous and heterozygous dominant genotypes is often a matter of how the phenotype is examined. Increasingly sophisticated biochemical tests are making it possible to distinguish phenotypically between the two dominant genotypes.

An interesting illustration of this recent phenomenon is provided by a re-examination of Mendel's round and wrinkled peas. When considered at the gross morphological level, seeds are clearly either round or wrinkled in form, with roundness dominant to wrinkledness. But, when seeds are examined microscopically or biochemically, a different and less obvious picture emerges (Box 3.2). It is, therefore, perhaps a good thing that neither sophisticated microscopes nor biochemical techniques were available to Gregor Mendel when he was performing his momentous experiments!

Observations such as those with the garden pea necessitate a redefinition of dominance. Dominance should be seen as a property of a pair of alleles in relation to the particular attribute of the phenotype being examined. Phenotypes may have many different physical and biochemical attributes: complete dominance may be observed for some of these and not others. This blurring of dominance relationship between alleles may at first

Box 3.2 Mendel's round and wrinkled peas

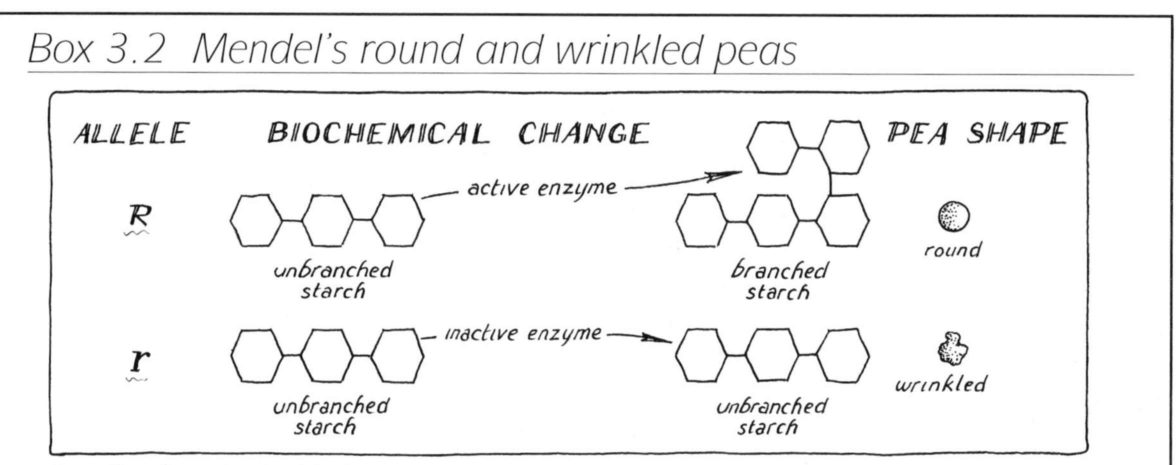

Formation of round and wrinked peas.

In 1991 a team of British geneticists identified the pea shape gene and worked out the physiological basis of roundness and wrinkledness. Starch can exist in two different forms – unbranched or **amylose** and highly branched or **amylopectin**. The pea shape gene encodes an enzyme known as **SBE1** (starch branching enzyme 1), which catalyses the conversion of amylose to amylopectin. The dominant allele **R** causes the formation of an active SBE1 enzyme. As a result the seed of an **RR** homozygote is full of the branched amylopectin, which enables the pea to maintain a rounded shape and to shrink uniformly as it ripens. In contrast, the product of the **r** allele is a non-effective enzyme. The seed of an **rr** homozygote instead contains amylose. It is irregular in shape and shrinks unevenly as it ripens. The result is a wrinkled seed. Although the heterozygote, **Rr**, has reduced levels of the active SBE1 enzyme, it is sufficient to convert most of the amylose to amylopectin. Thus, the amylopectin content is high enough to result in uniform shrinking and so a round seed.

seem confusing. It does not, however, detract from the useful conclusions that can be drawn from analysing the phenotypic ratios among the progeny of crosses. A 1 : 2 : 1 or 3 : 1 phenotypic ratio among offspring indicates the same thing – that one gene with two alleles is determining a given trait, and that a cross has been made between two heterozygotes. The different ratios then tell us whether or not the heterozygote can be distinguished phenotypically from the dominant homozygote, i.e. whether the two alleles show complete or incomplete dominance.

3.6 Codominance

Rather than a heterozygote showing an intermediate phenotype to either homozygote, it is possible for its expression to be additive, i.e. both alleles make an equal contribution to the phenotype so that the heterozygote shows distinctive features of both homozygous traits. This situation is described as **codominance** and is most commonly observed with biochemical traits. A good, and often quoted, example of codominance is the relationship between two of the three alleles that determine the human ABO blood groups.

The ABO blood groups result from a variation in two different polysaccharides, which we can call **A** and **B**, present on the surface of red blood cells. The ABO blood group gene encodes an enzyme needed for synthesis of these polysaccharides. Polysaccharide **A** is synthesized by an enzyme encoded by allele $\mathbf{I^A}$ and polysaccharide **B** is synthesized by an enzyme encoded by allele $\mathbf{I^B}$. A third allele, $\mathbf{I^i}$, produces no functional enzyme. Thus, the gene determining the ABO blood groups also shows multiple allelism (Section 3.1). Table 3.3 shows the relationship between the various genotypes and phenotypes.

Table 3.3 The genetic control of human ABO blood groups

Genotype	Enzyme produced	Polysaccharide present	Blood group
I^AI^A	Type A	A	A
I^AI^i	Type A	A	A
I^BI^B	Type B	B	B
I^BI^i	Type B	B	B
I^AI^B	**Types A and B**	**A and B**	**AB**
I^iI^i	none	none	O

The AB blood group is produced in individuals of heterozygous genotype **AB**. Each allele is producing a polysaccharide synthesizing enzyme, and so two different polysaccharides, **A** and **B**, are produced. In the **AB** heterozygote, therefore, both alleles are contributing equally to the phenotype. These alleles are described as **codominant**. Note that a 'normal' dominant/recessive relationship exists between alleles $\mathbf{I^A}$ and $\mathbf{I^i}$; and also between alleles $\mathbf{I^B}$ and $\mathbf{I^i}$. Blood group O is the recessive phenotype, only possible when individuals are homozygous recessive.

Knowing the dominance relationships between the three alleles determining the ABO blood groups means that predictions about genotypes and phenotypes can be made in the normal Mendelian way. Examples are given in the problems at the end of the chapter. One example is, however, considered in Fig. 3.6, because it illustrates a useful Mendelian ratio. Figure 3.6 shows the blood groups expected among children of parents who are both blood group AB (a rare situation in the UK, about 1 in 1000 parents). As with situations of incomplete dominance, if two alleles are codominant a 1 : 2 : 1 phenotypic ratio would be expected among the progeny when heterozygotes mate.

Parental cross **AB** **AB**

I^AI^B I^AI^B

Possible gametes I^A I^B I^A I^B

Gametes	I^A	I^B
I^A	I^AI^A blood group A	I^AI^B blood group AB
I^B	I^AI^B blood group AB	I^BI^B blood group B

Phenotypic ratio of blood groups among children: 1 A : 2 AB : 1 B

Fig. 3.6
Predicting ABO blood groups.

3.7 Lethal alleles

There are certain genes whose products are essential for life. Any allele that codes for a faulty product will therefore be lethal in the homozygous state. An individual heterozygous for a lethal allele generally shows an affected, but less severe, phenotype. One striking example of the effect of a lethal allele is the cat with no tail – the Manx cat (Fig. 3.7). All Manx cats are heterozygotes. In the homozygous state the tail-less allele is believed to cause such extreme spinal defects as to be non-conducive to life: homozygous embryos are reabsorbed by the mother cat.

Because Manx cats are heterozygotes, we might initially expect a 3 : 1 phenotypic ratio amongst the offspring of a mating between two Manx cats. Breeders of Manx cats, however, report a different outcome: smaller than average litter sizes with a 2 : 1 phenotypic ratio among the kittens, i.e. twice as many tail-less to tailed kittens. Figure 3.8 explains this result. One resulting genotype is lethal and so an unusual 2 : 1 phenotypic ratio is observed among the surviving progeny.

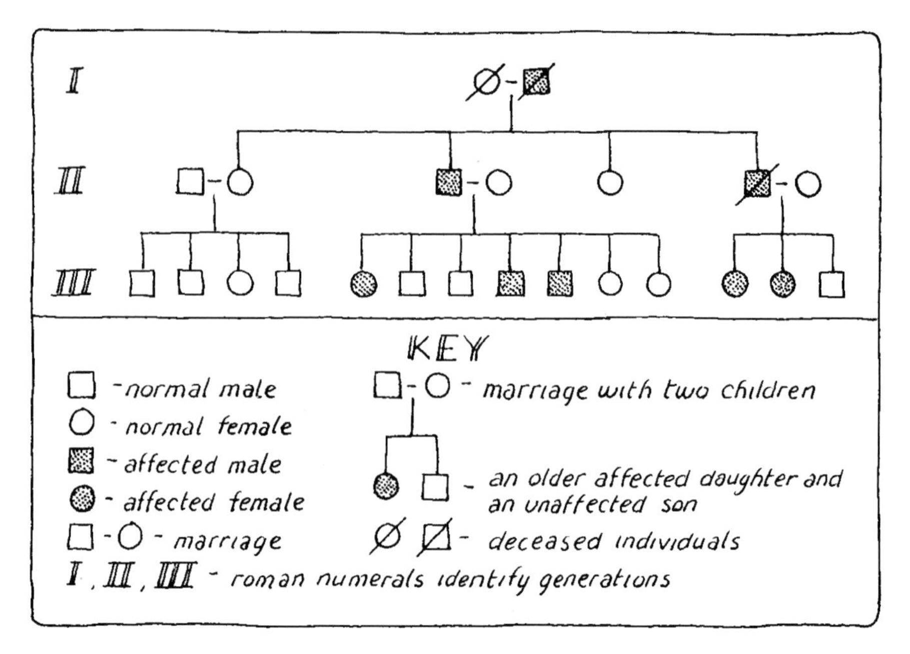

Fig. 3.9

Pedigree illustrating the inheritance of Huntingdon's disease.

The pedigree in Fig. 3.9 shows that the trait:

- affects both sexes;
- appears in each generation within an affected family;
- affects about half of all the children in an affected family.

These are characteristic segregation features for a condition caused by the presence of a dominant allele at a single locus. Affected individuals are heterozygotes, **Hh**, and non-affected people are homozygous normal, **hh**. The possession of two dominant alleles is inhibitory to normal development. For individuals with one dominant allele, symptoms (involuntary movements and progressive central nervous system degeneration) do not generally begin to show until they are in their thirties, with death following 15 to 20 years later. Because possession of this allele is ultimately fatal we recognize it as a **dominant lethal allele**. Although the allele is common within the family represented in Fig. 3.9, it is important to remember that the dominant allele causing Huntington's disease is very rare; an average of 1 in 10,000 individuals possess the allele.

Figure 3.10 shows a typical pedigree segregation pattern for a recessive trait, in this case the disease that causes dysfunction of the lungs, pancreas and digestive system – cystic fibrosis. The condition:

- affects both sexes;
- is not observed in each generation;
- has a low frequency of expression (especially when compared to the dominant trait in Fig. 3.9).

The pedigree in Fig. 3.10 illustrates another characteristic feature of recessive traits, particularly rare ones: parents of affected individuals are often related. The parents of IV-1 to IV-3 are first cousins. They are also

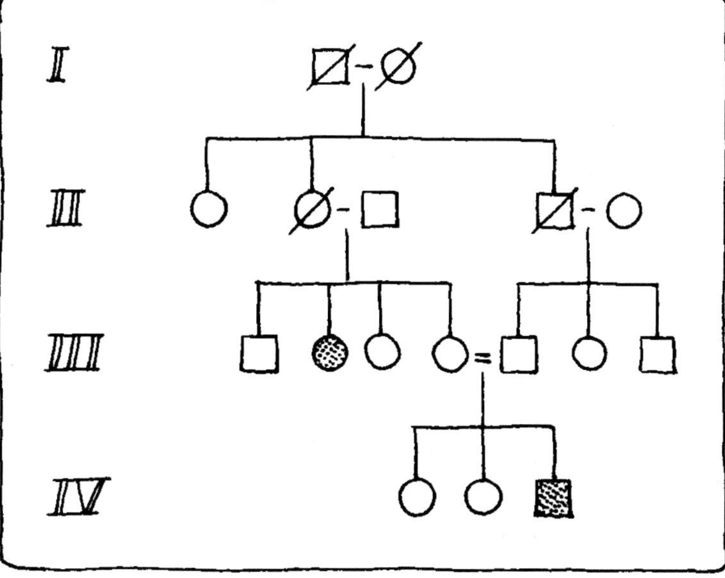

Fig. 3.10

Pedigree illustrating the inheritance of cystic fibrosis.

heterozygotes that each passed their recessive allele onto IV-3. The recessive cystic fibrosis allele must have been 'silently' passed from parent to children for many generations. It is only when two heterozygotes, or **carriers**, have children together that there is the chance of a homozygous expressing state. Each child of these carrier parents (III-4 and III-5) had a 1 in 4 chance of expressing the recessive condition. This can be demonstrated by drawing out the cross as we have been doing before (Fig. 3.11).

The Punnett Square in Fig. 3.11 shows the possible genotypes and phenotypes when two heterozygous individuals have children. They have a 1 in 4 chance of conceiving a child expressing the recessive phenotype. It

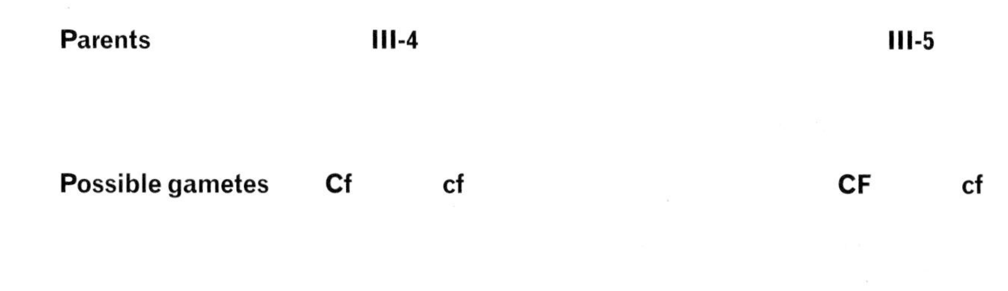

Gametes	CF	cf
CF	**CFCF** normal	**CFcf** normal
cf	**CFcf** normal	**cfcf** cystic fibrosis

The probability of a child born to heterozygous parents suffering from cystic fibrosis is 1 in 4

Fig. 3.11

Predicting the probability of expressing cystic fibrosis (cystic fibrosis results from a defect in a protein controlling transmembrane chloride transport. **CF** represents the normal transporter allele and **cf** the allele producing the faulty protein).

does not, however, mean that we can expect only one out of every four children to have cystic fibrosis. If their first child suffers from cystic fibrosis it is no guarantee that a couple's three subsequent children will be unaffected. Each fertilization is an independent event. A Punnett Square shows the possible outcomes of a single fertilization. It is possible for two carriers to have four children and each time gametes carrying the recessive allele to fertilize, so that all four children have the disease. The probability of this occurring is admittedly low, but it is certainly possible. We can calculate the possibility of this situation using the **multiplication rule**.

3.10 The multiplication rule

The Punnett Square of Fig. 3.11 shows the probability of any conceived child having a given genotype and phenotype. We want to predict the probability of four successive children having the same genotype and phenotype, i.e. to be a recessive homozygote and suffer from cystic fibrosis.

Each fertilization is an independent event with a given probability. The probability that the same event occurs in successive fertilizations is calculated by **multiplying together the probability of each independent event**. Thus, in the example under discussion:

The probability that a child expresses cystic fibrosis is $\frac{1}{4}$

The probability that four successive children express cystic fibrosis is $\frac{1}{4} \times \frac{1}{4} \times \frac{1}{4} \times \frac{1}{4} = \frac{1}{256}$

The multiplication rule enables us to define any sequence of events and to work out their probability. We might, for example, be interested to know what the probability is of a first child suffering from cystic fibrosis and a second being disease-free. Referring again to the Punnett Square of Fig. 3.11:

The probability of a child expressing cystic fibrosis is $\frac{1}{4}$

The probability of a child being disease-free is $\frac{3}{4}$

The probability of the first child expressing cystic fibrosis and the second child being disease-free is $\frac{1}{4} \times \frac{3}{4} = \frac{3}{16}$

We could have asked a more general question. A couple know that they are both carriers for the cystic fibrosis allele. They want to know the probability, if they have two children, of one having the disease. This could be achieved in one of two ways:

1. First child has cystic fibrosis; second child disease-free
2. First child disease-free: second child has cystic fibrosis

The probability of each of the above two events is:

1. $\frac{1}{4} \times \frac{3}{4} = \frac{3}{16}$
2. $\frac{3}{4} \times \frac{1}{4} = \frac{3}{16}$

Because either of the two scenarios is possible, they both have to be taken into consideration when calculating a final probability, so now the **addition rule** applies:

Probability of one of two children having cystic fibrosis
= probability of pattern 1 + probability of pattern 2
$= {}^{3}/_{16} + {}^{3}/_{16} = {}^{6}/_{16}$

Thus, the answer to the couple's question about the probability of one of two children having cystic fibrosis is ${}^{6}/_{16}$.

It needs practice with a variety of different situations to confidently know when to multiply and when to add probabilities. Some examples are included in the problems at the end of the chapter.

Summary

- Many alleles can exist at each gene locus. One individual, however, can possess only two alleles, so the inheritance of multiple alleles conforms to Mendel's Law of Segregation.
- Alleles at a locus may not show a clear-cut dominant/recessive relationship. Dominance relationships between alleles often need to be carefully defined in terms of the particular aspect of the phenotype being examined, e.g. physical or behavioural or biochemical.
- Only if there is no detectable difference between a homozygote and a heterozygote can an allele be described as completely dominant. Otherwise we recognize incomplete dominance and codominance.
- A 1 : 2 : 1 phenotypic and genotypic ratio in a cross between two heterozygotes indicates alleles are incompletely dominant or codominant.
- A 2 : 1 phenotypic ratio between two heterozygotes indicates an allele is lethal as a homozygote.
- Distinctive segregation patterns in human pedigrees can indicate whether a specific condition results from a dominant or recessive allele at a given locus.

Problems

1. In the four-o'clock plant the allele for red flowers is incompletely dominant over the allele for white ones, so the heterozygotes are pink. What ratio of flower colours would you expect among the offspring of the following crosses?
   (a) pink × pink-flowering plants
   (b) white × pink-flowering plants
   (c) white × red-flowering plants
   (d) white × white-flowering plants

2. Henrik is blood group B, like his mother. His older sister is group A and his younger brother is group O. What are the genotypes of his parents?

3. Three alleles at a single locus determine whether guinea pig coat colour is yellow, cream or white. One homozygote, $\mathbf{C^YC^Y}$, is yellow, the other homozygote is white, $\mathbf{C^wC^w}$, while the heterozygote, $\mathbf{C^WC^Y}$, is cream. If two cream-coloured individuals are mated what phenotypes will be observed among the progeny, and in what ratios?

4. The shape of radishes may be long, $\mathbf{S^LS^L}$, round, $\mathbf{S^RS^R}$, or oval, $\mathbf{S^LS^R}$. If plants producing long radishes are crossed with some producing round ones, and the resulting F_1 plants selfed, will any oval-producing plants be present in the F_2 generation?

5. In 1943 actress Joan Barry sued Charlie Chaplin for support of a child that she claimed he had fathered. She was blood group A, Chaplin group O and the child group B. Chaplin was deemed the father and ordered by the court to pay maintenance for the child. Do you agree with the court's decision?

6. In chickens the dominant allele (**C**) produces 'creepers' (short-legged chickens) when it is heterozygous, but when homozygous the offspring never hatch. A creeper cockerel and hen are mated and produce 21 viable offspring. How many do you expect to have normal legs?

7. It was wondered whether migratory behaviour had a genetic basis. A system was developed for studying migration in the laboratory. It was observed that birds which migrated became much more energetic at the appropriate times of the year. Sensors were therefore fitted to their perches and the number of jumping movements, measured electronically, correlated with migratory behaviour. Crosses were made between birds from two populations of the blackcap warbler: a non-migratory one that lived all year in Africa and another that spent the summers in Germany and the winters in Africa. All the resulting F_1 birds showed jumping movements intermediate in number between the two parents. Suggest
   (a) what the genetic basis of migratory behaviour is
   (b) what the result would be of crossing the F_1 birds with birds from the non-migratory African population.

8. One gene with five alleles controls the patterning on lentil seeds. Two of the five alleles produce two different marbled patterns ($\mathbf{L^{M1}}$ and $\mathbf{L^{M2}}$). The other three alleles result in spotted ($\mathbf{L^S}$), dotted ($\mathbf{L^D}$) or clear ($\mathbf{L^C}$) seeds. The dominance relationship is in the order in which these alleles have been presented, i.e. $\mathbf{L^{M1}}$ is dominant to the other four, then $\mathbf{L^{M2}}$, $\mathbf{L^S}$, $\mathbf{L^D}$ and finally $\mathbf{L^C}$ is recessive to all others. Thus, what would be the expected phenotypic ratios among the plants of the following crosses:
   (a) marbled-1 ($\mathbf{L^{M1}L^{M1}}$) × dotted ($\mathbf{L^SL^S}$)
   (b) marbled-2 ($\mathbf{L^{M2}L^C}$) × dotted ($\mathbf{L^DL^C}$)
   (c) marbled-1 ($\mathbf{L^{M1}L^D}$) × marbled-2 ($\mathbf{L^{M2}L^D}$)

9. In foxes two alleles exist at a particular locus, **P** and **p**. The homozygote, **PP**, results in lethality, the heterozygote produces a platinum coat while the homozygote, **pp**, produces a silver coat. What phenotypes, and in what ratios, are obtained when platinum foxes breed?

10. Horses can be cremello (a light cream colour), chestnut (a deep brown) or palomino (golden with a white tail and mane). Cremello and chestnut horses always breed true, but palominos never do. From the results below, deduce the mode of inheritance of these colours and complete the table below, showing the results of matings between cremello, chestnut and palomino horses.

Parents	F_1 progeny	Parents' genotypes
(a) cremello × palomino	50% cremello; 50% palomino	
(b) chestnut × palomino	50% chestnut; 50% palomino	
(c) palomino × palomino	25% cremello; 50% palomino: 25% palomino	

11. Overleaf is a pedigree for an inherited form of deafness. (Affected individuals are represented by a shaded symbol.)
   (a) Is the mode of inheritance of this trait recessive or dominant?
   (b) Choose suitable systems and give the genotypes of individuals (i) II-1; (ii) II-3; (iii) III-2 and (iv) IV-3.

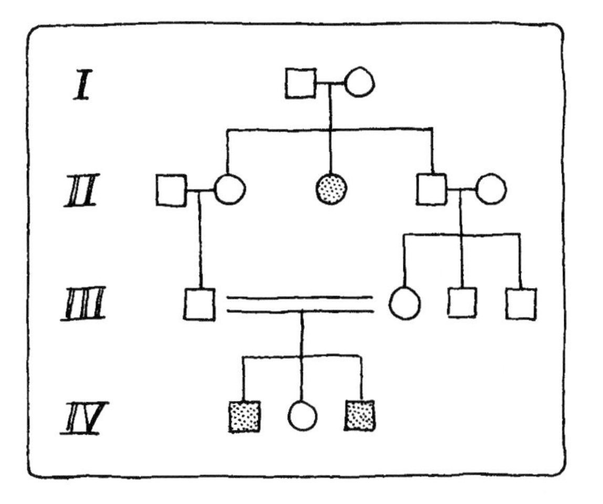

12. In pigs the normal cloven-footed condition is dominant to mule-footed. A true-breeding cloven-footed pig is crossed to a mule-footed one. The resulting F_1 pigs are self-fertilized. What are the following probabilities for the F_2 generations:

(a) that the first pig examined is mule-footed?

(b) that the first pig examined is mule- or cloven-footed?

(c) that the first three pigs examined are cloven-footed?

13. In fowl, the alleles for black and white feathers are co-dominant. The heterozygote, therefore, produces both black and white feathers and the resulting colour is 'blue' or 'erminette'. If an erminette cockerel mates with an erminette hen, what is the probability that from a clutch of four eggs:

(a) all will hatch into erminette fowls

(b) all will hatch into white fowls

(c) the first egg will hatch into a white fowl and the other three produce black ones?

14. Schilder's disease is a progressive degeneration of the central nervous system in humans, which leads to death at age two. It is inherited as a recessive trait. A couple lose their first child to the disease. They would very much like a family of two children. What is the probability that two further conceptions will produce disease-free children?

4 Dihybrid inheritance

In the previous two chapters we have been considering traits determined by allelic variation at a single locus. Such monohybrid traits are, however, in the minority. The vast majority of traits in all organisms arise from the action of two or more, often many, genes. In addition interactions between genes and the environment can be important in determining the final phenotype. The outcome of controlled breeding experiments and analysis of ratios among progeny can still provide important clues to the genetic basis of such traits. This chapter focuses on various ways in which two genes can interact to determine a single trait. Identifying these situations requires an understanding of the principles of **dihybrid inheritance**, i.e. crosses that simultaneously consider the inheritance of alleles at two separate loci. Gregor Mendel first performed these kinds of crosses: he was interested in examining patterns of inheritance of two different traits, each controlled by a single gene. He discovered that the F_2 generation showed some characteristic ratios, which remain extremely useful in flagging up the involvement of two genes in the expression of a trait.

This chapter introduces the expected patterns of phenotypic expression in the F_1 and F_2 generations when:

- two genes control the expression of two separate characteristics;
- two genes control the expression of a single trait.

4.1 Two genes – two characters

One summer a tomato grower planted some purple-leaved and yellow-fruiting tomato plants in his herbaceous border. These unusual plants attracted a lot of attention and so he decided that he would like to start selling them on a commercial basis. He already understood the genetic basis of yellow-fruiting tomatoes – that it was the result of a recessive allele at a single locus (see Chapter 2). He was, however, unaware of the genetic basis of leaf colour and how or whether it was possible to consistently breed his desired combination of characteristics. The tomato grower was in luck. Mendelian principles could be applied to ensure a constant supply of the desired phenotype.

To understand the principles involved in selectively breeding plants with the desired two characters, let us return to the standard procedure used in Chapter 2 when establishing the principles of monohybrid inheritance, i.e. that of crossing pure-breeding parents to achieve an F_1 generation that is then crossed among itself to produce the F_2 generation. In this current example two characteristics are under consideration:

- fruit colour: either red or yellow
- leaf colour: either green or purple

Suppose pure-breeding, yellow-fruiting, purple-leaved plants are crossed with pure-breeding, red-fruiting, green-leaved plants. The resulting phenotypes of the F_1 and F_2 plants are shown in Fig. 4.1.

Parental cross pure-breeding, yellow-fruiting, purple-leaved × pure-breeding red-fruiting, green-leaved

F_1 result red-fruiting, green-leaved

F_2 result	red-fruiting green-leaved	red-fruiting purple-leaved	yellow-fruiting green-leaved	yellow-fruiting purple-leaved
	9	3	3	1

Fig. 4.1
Investigating the inheritance of tomato fruit and leaf colour.

As with monohybrid examples, a striking pattern of inheritance can be seen: certain phenotypes have disappeared in the F_1 generation, but reappeared again in the F_2. Indeed, four different phenotypes are observed in the F_2 generation, representing all possible combinations of the individual characters. If this experiment is repeated enough times, a consistent relationship between the different F_2 phenotypes emerges, i.e. the four different phenotypes of red-fruiting green-leaved, red-fruiting purple-leaved, yellow-fruiting green-leaved and yellow-fruiting purple-leaved plants occur in a recognizable ratio of 9 : 3 : 3 : 1.

Taking our cues from the monohybrid work we can conclude that the phenotypes expressed in the F_1 generation are the dominant ones, here red-fruiting and green-leaved. We can therefore rewrite this current cross, using symbols for the different alleles:

Let **R** = allele for red fruiting

Let **r** = allele for yellow fruiting

Let **G** = allele for green leaves

Let **g** = allele for purple leaves

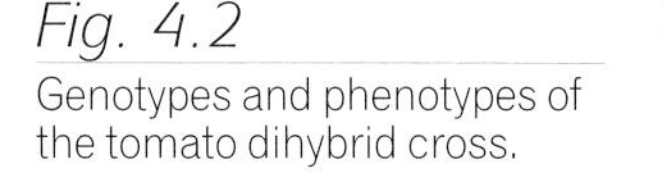

Fig. 4.2

Genotypes and phenotypes of the tomato dihybrid cross.

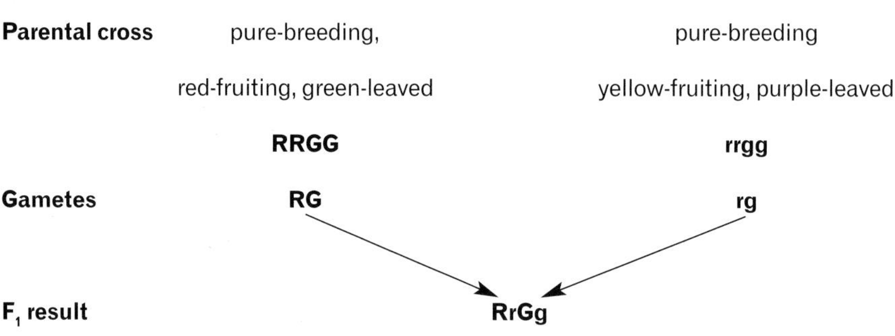

F_2 result

Gametes	**RG**	**rG**	**Rg**	**rg**
RG	**RRGG** red green	**RrGG** red green	**RRGg** red green	**RrGg** red green
rG	**RrGG** red green	**rrGG** yellow green	**RrGg** red green	**rrGg** yellow green
Rg	**RRGg** red green	**RrGg** red green	**RRgg** red purple	**Rrgg** red purple
rg	**RrGg** red green	**rrGg** yellow green	**Rrgg** red purple	**rrgg** yellow purple

red-fruiting green-leaved	red-fruiting purple-leaved	yellow-fruiting green-leaved	yellow-fruiting purple-leaved
R-G-	**R-gg**	**rrG-**	**rrgg**
9	3	3	1

There are several things to note from the above representation of the tomato dihybrid cross:

1. Each gamete has one allele for fruit colour and one for leaf colour.
2. There are four different types of gametes possible for each F_1 heterozygote: each fruit colour allele can be found in a gamete with either of the two leaf colour alleles.
3. Because there are four different gametes possible for each F_1 male and also for each F_1 female, this means that there are 16 possible fertilization outcomes to produce the F_2 generation.
4. These 16 possible fertilizations result in nine different genotypes determining the four different F_2 phenotypes (Table 4.1).

Genotype	Phenotypes
RRGG RrGG RRGg RrGg	Red-fruiting, green-leaved
RRgg Rrgg	Red-fruiting, purple-leaved
rrGG rrGg	Yellow-fruiting, green-leaved
rrgg	Yellow-fruiting, purple-leaved

Table 4.1
A summary of F_2 genotypes and corresponding phenotypes

5. These four different F_2 phenotypes occur in a ratio of 9 : 3 : 3 : 1, with the different classes distributed as shown in Table 4.2 with regards to dominance and recessiveness.

Fertilization outcome of F_1 cross	Phenotypic expression of trait 1	Phenotypic expression of trait 2
9/16	Dominant	Dominant
3/16	Dominant	Recessive
3/16	Recessive	Dominant
1/16	Recessive	Recessive

Table 4.2
Dominant/recessive distribution among the F_2 generation in a dihybrid cross

6. If we consider each trait separately, i.e. just fruit or leaf colour, then it can be seen that each trait is inherited in the F_2 generation in the 3 : 1 ratio predicted by Gregor Mendel's Law of Segregation. In the Punnett Square (Fig. 4.2) there are twelve red fruiters to four yellow ones and, likewise, twelve green-leaved plants to four purple-leaved ones. This shows that the inheritance of the alleles of one gene is unaffected by the inheritance of alleles at the other locus. Mendel realized this independence and encapsulated it in his second law – **The Law of Independent Assortment**. Using modern genetic terminology this can be stated as:

   Each pair of alleles segregates independently so that in the gametes one member of each pair is equally likely to appear with either of the two alleles of the other pair.

4.2 Test cross

Our tomato grower was delighted to know how to guarantee production of his yellow-fruiting purple-leaved plants. Any such plant was homozygous recessive, and therefore pure-breeding. He also wanted to ensure a supply of red-fruiting green-leaved plants for commercial tomato production. To this end he had discovered that he also needed pure-breeding plants. In the F_2 generation above, the red-fruiting green-leaved plants can be one of four different genotypes (**RRGG**, **RrGG**, **RRGg** and **RrGg**). To distinguish between these, and so identify the pure-breeding plant (**RRGG**), the tomato grower needs to do some test crosses. You may remember from Chapter 2 that this involves crossing plants of dominant phenotype, but unknown genotype, with homozygous recessive individuals. The phenotypes among the progeny indicate the genotypes of the phenotypically dominant plants. In this example each F_2 red-fruiting green-leaved plant needs to be crossed with a double homozygous recessive plant (**rrgg**), which will be yellow-fruiting and purple-leaved. The possible outcomes, depending upon the genotypes of the red-fruiting green-leaved plant, are shown in Fig. 4.3. Thus, to ensure a guaranteed supply of red-fruiting green-leaved plants the tomato grower needs to keep seeds from plants yielding 100% red-fruiting green-leaved plants when crossed with the double recessive.

This demonstration of the outcomes of various test crosses has also identified another useful Mendelian ratio. If a cross of a phenotypically dominant individual with a recessive homozygote yields progeny equally distributed among four phenotypic classes in a 1 : 1 : 1 : 1 ratio, this indicates that the individual of unknown genotype is a double heterozygote, here **RrGg**.

4.3 Mid-chapter summary

The previous two sections have established some important principles relating to the simultaneous inheritance of two genes, where each gene determines a different trait, i.e.

Gene A → trait 1

Gene B → trait 2

It is not, as stated earlier, uncommon to find a single characteristic under the influence of two genes, i.e.

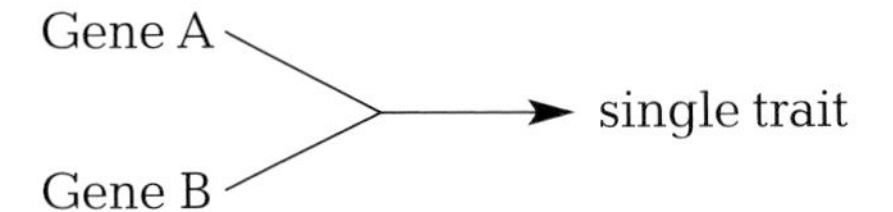

The principles established from looking at patterns of dihybrid inheritance, when two traits are involved, can be usefully applied to the second situation to enable various relationships between alleles at two different loci to be deduced. Clues come from looking at the ratios of phenotypic types among the progeny in the F_2 generation of a standard Mendelian cross.

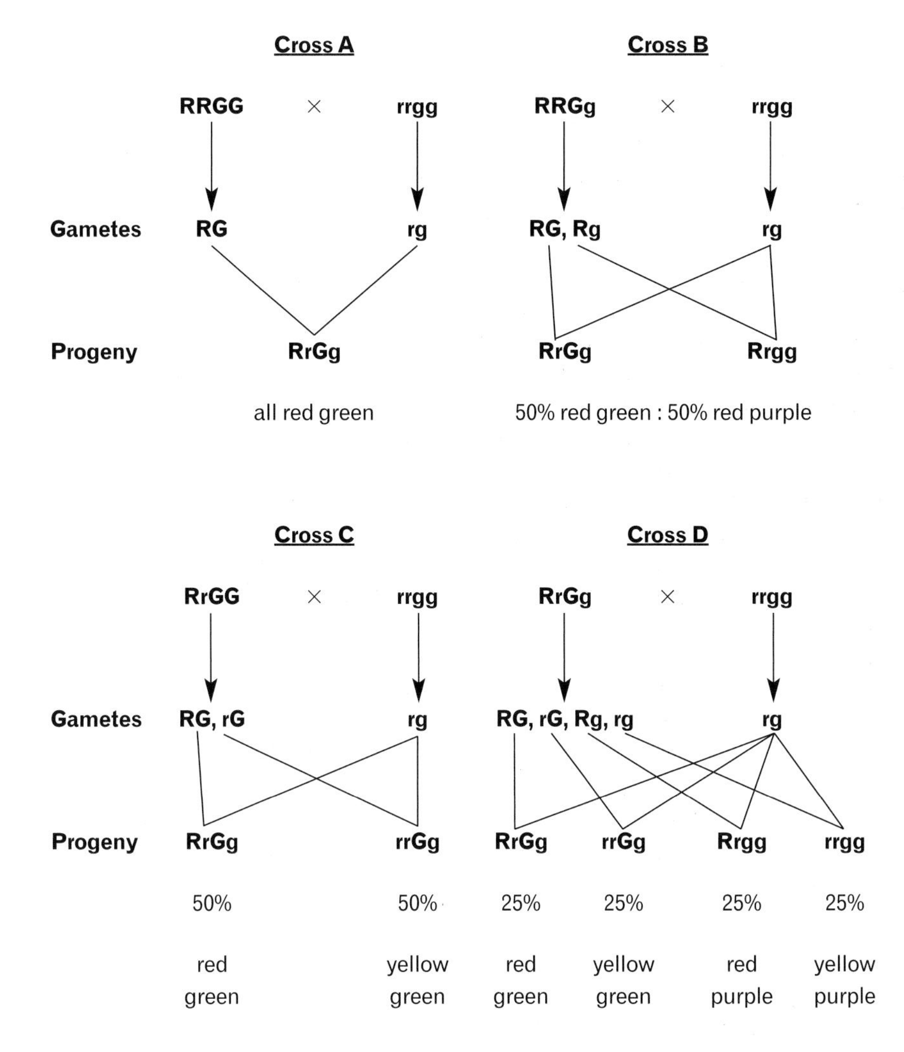

Fig. 4.3

Possible outcomes of test crossing an F_2 red-fruiting green-leaved plant.

Two independently assorting genes typically produce a 9 : 3 : 3 : 1 ratio among four different phenotypes. Interactions between the alleles at two loci can result in modifications to this standard ratio. Different modifications indicate different kinds of interactions between the genes. These will be discussed in the following sections, although Section 4.4 deals with an example where the 9 : 3 : 3 : 1 F_2 ratio is maintained.

4.4 Two genes – one trait

One of the first pieces of evidence that a trait can be influenced by more than one gene was obtained by William Bateson (Fig. 4.4) and Reginald Punnett (Fig. 2.5) working with poultry during the early 1900s, shortly after the rediscovery of Mendel's work. Different breeds of domestic fowl have different shaped combs (Fig. 4.5). In a now classical experiment, Bateson and Punnett crossed rose-combed Wyandottes with pea-combed Brahmas. All their F_1 chickens possessed a novel comb shape – walnut. When these walnuts were interbred, four different comb shapes were observed among the resulting F_2 progeny: rose, pea, walnut and the distinctive spiked single comb already known in the Leghorn breed (Fig. 4.6). The numerical relationships among the F_2 offspring immediately gave clues to the genetic basis of comb shape. The 9 : 3 : 3 : 1 ratio of walnut, pea, rose and single-combed fowl indicated that two independently assorting genes controlled expression of comb shape.

The two genes controlling comb shape are designated **R** and **P**. Table 4.3 shows the relationship between the different genotypes and the four different comb shapes, and Fig. 4.7 represents Bateson and Punnett's original experiment.

Table 4.3
Relationships between genotypes and comb shape

Genotype	Phenotype
P-R-	Walnut (double dominant)
pprr	Single (double recessive)
P-rr	Pea (one locus dominant)
RRp-	Rose (other locus dominant)

Fig. 4.4
William Bateson is often credited with inventing the language of genetics; for example, allelomorph (later shortened to allele), zygote, homozygote and heterozygote. In 1905, noting that 'a word is badly wanted' to describe the study of inheritance and variation, he suggested 'genetics might do'.

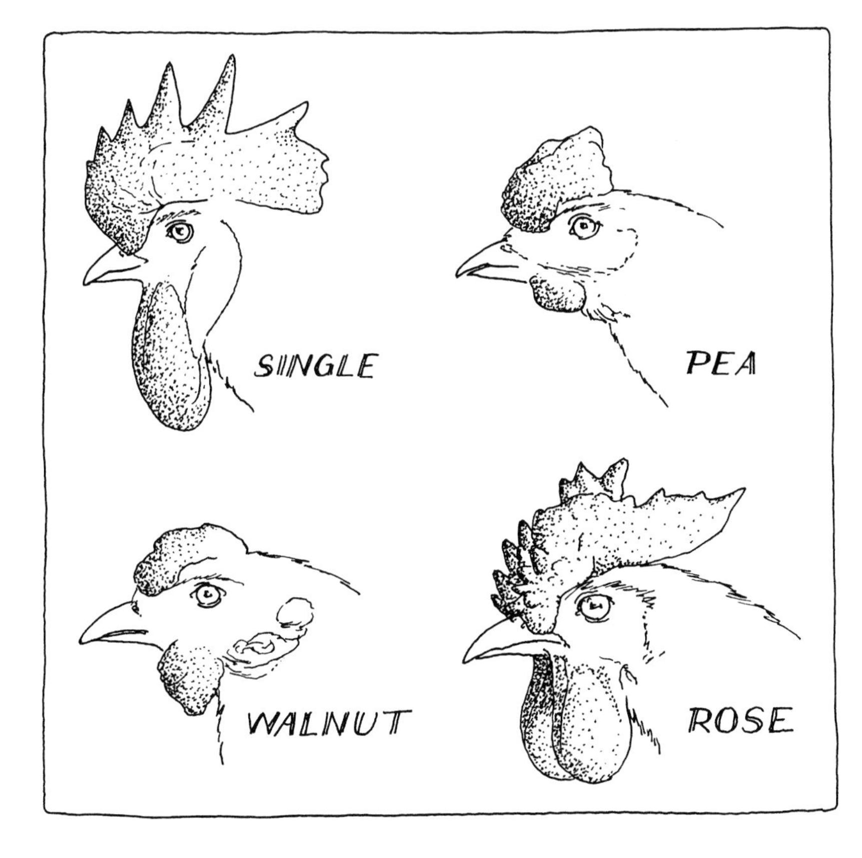

Fig. 4.5
Fowl comb shapes.

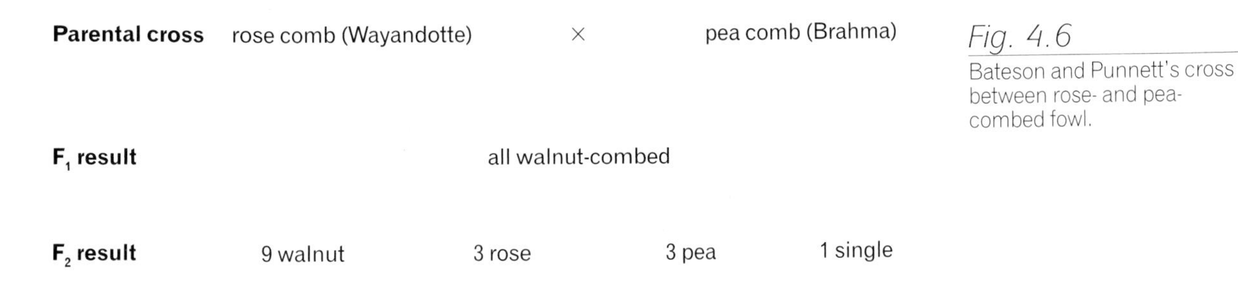

Fig. 4.6
Bateson and Punnett's cross between rose- and pea-combed fowl.

This section has described a situation where the products of two independently assorting genes interact to influence expression of a single trait. Each genotype class (**P-R-**, **P-rr**, **ppR-** and **pprr**) determines a different phenotype. This is not always the case. Sometimes three, or even just two, phenotypes are observed in the F_2 generation in ratios such as 9 : 3 : 4 or 15 : 1. We know, however, that we are still dealing with two independently assorting genes as the F_2 ratio is clearly a modification of the standard 9 : 3 : 3 : 1. The observed phenotypes are in proportions of 16. From the way the 9 : 3 : 3 : 1 ratio is modified it becomes possible to deduce something about the nature of the genes' interactions.

Table 4.8
Calculating a value for χ^2

Phenotype	Observed numbers (*O*)	Expected numbers (*E*)	*O* – *E*	$(O-E)^2$	$\frac{(O-E)^2}{E}$
Red fruits, green leaves	135	150	–15	225	1.5
Red fruits, purple leaves	163	150	13	169	1.13
Yellow fruits, green leaves	155	150	5	25	0.17
Yellow fruits, purple leaves	148	150	–2	4	0.03
Total	600	600			**2.83**

Null hypothesis – there is no difference between observed results and an expected ratio of 1 : 1 : 1 : 1
Significance level = 0.05; degrees of freedom = 3; calculated χ^2 value = 2.83; critical χ^2 value = 7.82
Because the calculated χ^2 value is less than the critical χ^2 value, at a significance level of 0.05 and 3 degrees of freedom, the null hypothesis can be accepted. The results, therefore, resemble a 1 : 1 : 1 : 1 ratio, and we can conclude that the red-fruiting green-leaved plant was heterozygous.

Summary

- Dihybrid crosses investigate the patterns of inheritance when two genes are involved.
- The segregation and subsequent inheritance of alleles at each gene occur independently.
- When two genes control variation in two different characteristics, a phenotypic ratio of 9 : 3 : 3 : 1 is obtained in the F_2 generation.
- The 9 : 3 : 3 : 1 F_2 ratio may be modified when both genes control the same characteristic. The nature of the modified ratio indicates the type of interaction between the two loci.
- Complementation occurs when a functional gene product of both loci is required for full expression of a character.
- Epistasis refers to the situation when either the dominant or the recessive allele at one locus prevents expression at the second.
- A chi-squared statistical test assesses whether observed data matches predicted results.

Problems

1. In the tomato purple stems are dominant to green and the presence of hairs on the stems is dominant to their absence. A purple, hairy-stemmed plant was self-fertilized and produced the following offspring:

   133 purple hairy; 47 purple hairless; 45 green hairy; 15 green hairless. What was the genotype of the self-fertilized purple hairy-stemmed plant?

2. Fur colour in mice may be black (**B**) or brown (**b**), and their tails may be short (**S**) or long (**s**). What proportion of the progeny of a cross between two heterozygotes (**BbSs**) will have brown hair and short tails, if fur colour and tail length assort independently?

3. Two pure-breeding blind strains of crickets were crossed to each other. The F_1 crickets all showed normal sight. When these F_1 crickets were crossed among each other, the resulting F_2 progeny consisted of 178 crickets with normal sight and 142 blind crickets.

   (a) Explain the results.

   (b) What fraction of the F_2 generation would you expect to be pure-breeding for sightedness?

   (c) What phenotypes would you expect among the progeny, and in what proportions, if the F_1 crickets were crossed to a double homozygous recessive individual?

4. In Shorthorn cattle polled (hornless) is dominant to horned. Coat colour may be red, white or a mixture of red and white hairs referred to as roan; the two colours being determined by two separate alleles at a single locus. A roan horned bull was mated on four different occasions with a white polled cow. All the resulting calves were polled, but five were white and four were roan. What were the genotypes of the parents?

5. Removal of infected honey bee larvae from a hive is under genetic control. It involves (i) uncapping the cell of an infected larvae; (ii) removing the infected larvae. Successful uncapping and removal depends upon appropriate alleles being present at two loci:

	Uncapping	Removal
U-R-	✗	✗
uuR-	✓	✗
U-rr	✗	✗
uurr	✓	✓

   One hundred worker bees were taken at random from a hive and their uncapping and removal behaviour investigated. Only six bees successfully uncapped and removed infected larvae. A further 20 uncapped but did not remove larvae, while the remaining 74 exhibited no nest cleaning behaviour. Suggest genotypes for these 100 bees.

6. Two varieties of corn produce colourless aleurone layers in their seeds. When crossed, the resulting seeds all possessed a purple aleurone layer. The same colour also appeared in 270 of the 480 F_2 plants. How do you explain inheritance of a coloured aleurone layer in corn?

7. Cyanogenesis is an anti-predation mechanism in the white clover (*Trifolium repens*). It is regulated by two loci. The first (**Ac/ac**) ensures production of glucoside and the second (**Li/li**) encodes the enzyme linamarase; which catalyses the production of hydrogen cyanide from the glucoside.

   $$\text{precursor} \xrightarrow{\textbf{Ac/ac}} \text{glucoside} \xrightarrow{\textbf{Li/li}} \text{hydrogen cyanide}$$

   Thus, if the two loci are in the dominant state (**Ac-Li-**), leaves readily produce hydrogen cyanide when crushed. If only the **Ac** locus is producing a functional product cyanogenesis still occurs, but at a slow, spontaneous rate. The plant is acyanogenic if the first gene product is absent. From a cross involving plants heterozygous at both loci, what is the probability that any one of the resulting clover plants will be

   (a) cyanogenic; (b) acyanogenic; (c) show slow cyanogenesis?

8. In horses a trotting gait (**T**) is dominant to a pacing gait (**t**), and black coat (**B**) is dominant to chestnut (**b**). What are the genotypes and phenotypes of the parents which produce offspring in the following proportions: one-eighth chestnut pacers : one-eighth chestnut trotters : three-eighths black trotters : three-eighths black pacers?

9. Achondroplasia (an inherited form of dwarfism) and short-sightedness are both dominantly inherited human traits. A man with achondroplasia marries a woman who is short-sighted. If both individuals are heterozygous:

   (a) What is the probability that any child has achondroplasia or is short-sighted?

   (b) What is the probability that a child is both short-sighted and has achondroplasia?

10. A cat breeder crossed two white cats expecting all the kittens to be white. She was greatly surprised to obtain two tabby and a black kitten among the litter. Next time the white female came into season the breeder repeated the mating. Again the litter contained black and tabby kittens. A friend suggested that the two white cats might each possess a dominant colour-suppressing gene, and that, when the cats were mated, non-suppressing alleles had segregated to some of the kittens, enabling different colours to be expressed. To test this hypothesis the breeder did one final mating. Collectively the three litters had yielded 17 white, 5 tabby and 2 black kittens. Use a χ^2 test to decide whether the breeder's friend was correct.

The chromosomal basis of inheritance

The rediscovery of Mendel's work on the nature of inheritance in 1900 led to a great burst of genetic experimentation. It quickly became clear that Mendel's laws were valid for a wide variety of organisms. Geneticists also realized that the inheritance patterns of Mendel's factors paralleled the behaviour of the newly discovered chromosomes during sexual reproduction. As a consequence, the **chromosome theory of inheritance** was proposed – that chromosomes were the carriers of Mendel's factors, or 'genes', as they soon became known.

This chapter considers:

- the physical nature of chromosomes;
- the transmission of chromosomes from cell to cell and from generation to generation during the processes of mitosis and meiosis, respectively;
- the correlation between the segregation patterns of genes, as discussed in Chapters 2 to 4, and the behaviour of chromosomes during meiosis.

5.1 The structure of chromosomes

A chromosome is one long molecule of DNA (see Chapter 11) arranged within a framework of protein molecules. Chromosomes can only be seen as discrete structures during the two processes of nuclear division, mitosis and meiosis, when they become highly condensed. It is, therefore, only during mitosis and meiosis that it is possible to distinguish the size, shape and number of chromosomes within a cell. Figure 5.1 is a scanning electron micrograph of human chromosomes, showing that each chromosome is a double structure, composed of a pair of **chromatids**, held together at the **centromere**.

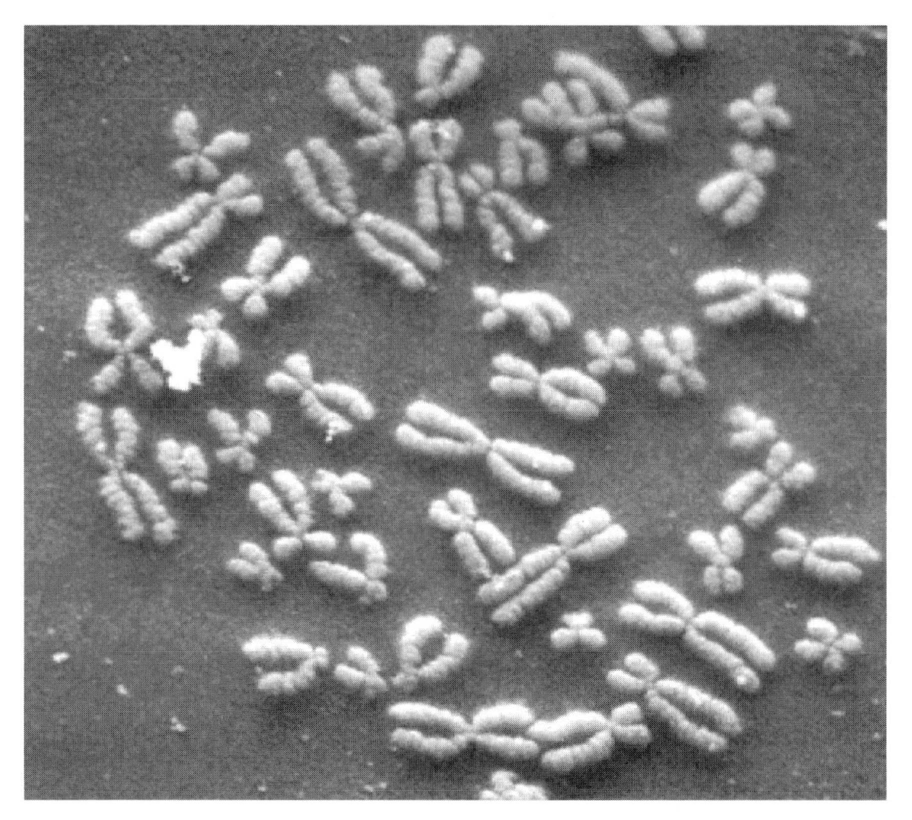

Fig. 5.1
Scanning electron micrograph of human chromosomes.

The frequently used terms 'chromosome' and 'chromatid' need to be distinguished. Within a normal working eukaryotic cell each chromosome is present as a single structure. When a cell divides it is essential that each new cell has a complete and accurate copy of an organism's genetic information. Thus, prior to division, each chromosome makes an exact copy of itself. During the early stages of both mitosis and meiosis the two new chromosomes remain attached at a site called the centromere, visible under the microscope as a constriction (Fig. 5.1). While the two replicas are attached they are known as chromatids. Once in separate nuclei they are again referred to as chromosomes.

Often a **karyotype**, a display of a cell's chromosomes ordered according to size, is made (Fig. 5.2). This enables a detailed analysis of a cell's chromosomes. A photograph is taken of a cell arrested in the middle of mitosis. The technique involves rupturing the nucleus, which results in a 'spread' of chromosomes (upper photograph, Fig. 5.2). A karyotype is produced from this first photograph by cutting out the individual chromosomes and arranging them in pairs according to size and shape. Most species show variation in the sizes of the different chromosomes. A group of similar-sized chromosomes, for example the chromosomes of group D in the human karyotype (Fig. 5.2), can be sorted into matching pairs because each different chromosome possesses a distinctive banding pattern when stained with chemical dyes (Fig. 5.2 and Box 5.1). Two chromosomes, the **sex chromosomes**, which determine the sex of an individual, are usually displayed separately at the end of a karyotype. This enables an individual's gender to be quickly recognized. Human females, for example, possess two identical **X chromosomes**, while males have one X and a much smaller **Y chromosome** (for more details see Chapter 6). All other chromosomes are known as **autosomes**.

Fig. 5.2
Karyotype of a G-banded human male. Autosomes are placed in six groups, A–G, according to size.

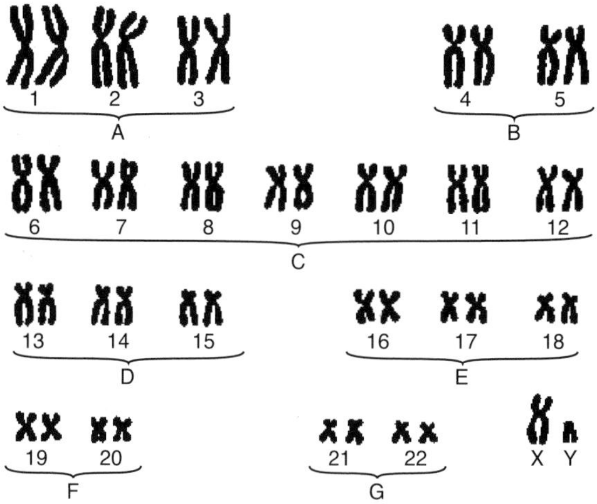

Box 5.1 Staining eukaryotic chromosomes

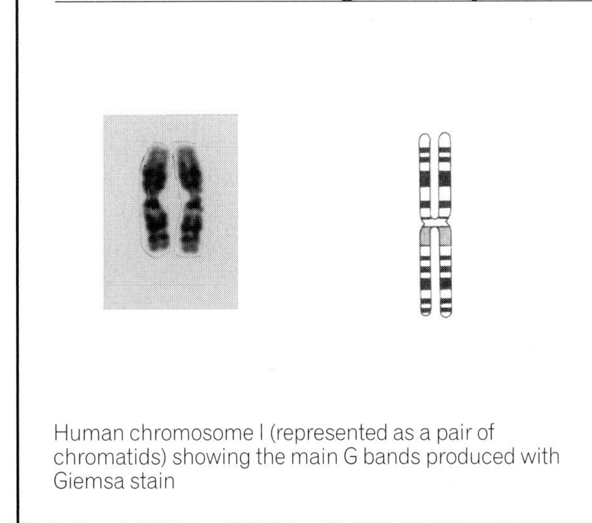

Human chromosome I (represented as a pair of chromatids) showing the main G bands produced with Giemsa stain

Until about 1970, mitotic chromosomes could only be distinguished under the light microscope on the basis of their relative size and the positions of their centromeres. Thus, the situation often arose that two or more chromosomes could not be separately identified. However, in about 1970, several different research groups developed techniques that produced differential staining along the long axis of chromosomes. The results were recognizable and reproducible banding patterns for the different chromosomes of a species. Several of the techniques used **Giemsa stain**. Depending upon the conditions associated with the use of the stain, different distinctive patterns are obtained, referred to as **G, C** and **R** banding. The variety of staining reactions, under different conditions, reflects the heterogeneity and complexity of chromosome composition.

5.2 The number of chromosomes

The karyotype in Fig. 5.2 shows that each human cell has 46 chromosomes. The number of chromosomes is constant for all cells of a given species, although it can vary enormously between species. For example, each cell of the scorpion possesses just four chromosomes, while some ferns, such as the adder's tongue, have more than 1000 chromosomes per cell! More common, however, is a chromosome number between 10 and 50 (Table 5.1).

Table 5.1
Chromosome number in a selection of animal and plant species

Animal		**Plant**	
Species name	**Chromosome number**	**Species name**	**Chromosome number**
Homo sapiens (human)	46	*Pisum sativum* (garden pea)	14
Pan troglodytes (chimpanzee)	48	*Solanum tuberosum* (potato)	48
Equus caballus (horse)	64	*Triticum aestivum* (wheat)	42
Felis domesticus (cat)	38	*Zea mays* (maize)	20
Oryctolagus cuniculus (rabbit)	44	*Vicia faba* (broad bean)	12
Xenopus laevis (toad)	36	*Gossypium hirsutum* (cotton)	56
Musca domesticus (housefly)	12	*Sequoiadendron giganteum* (redwood)	22
Mus musculus (mouse)	40	*Lycopersicon esculentum* (tomato)	24
Alligator missispiensis (alligator)	32	*Antirrhinum majus* (snapdragon)	8
Carssinus auratus (goldfish)	94	*Oryza sativa* (rice)	24

5.3 Homologous pairs of chromosomes

The chromosomes of any normal working somatic cell can generally be arranged in matching pairs (Fig. 5.2). These pairs are referred to as **homologous pairs**. The 46 chromosomes of a human somatic cell, therefore, represent two **sets** of chromosomes, each containing 23 different chromosomes. One of these sets originally came from the mother – the **maternal set** – and the other from the father – the **paternal set**.

The two chromosomes of a homologous pair have the same size, shape and functions. They have a characteristic set of genes that determine a collection of different traits. A particular gene will always be found at the same **locus**, or position, on a given chromosome and on its homologous partner. Much research is currently directed towards determining, or **mapping**, the precise locations of the 30,000 to 50,000 genes on the 23 pairs of different human chromosomes. Various other mapping projects are currently under way in a range of other species, for example the mouse, wheat, thale cress and fruit fly.

5.4 Diploid and haploid cells

Any cell containing two sets of chromosomes is described as **diploid** and sometimes represented as **2n**, where n represents the number of chromosomes of one set. Gametes generally have just one set of chromosomes. These are referred to as **haploid** cells, and represented as **n**. The genetic information present in a haploid set of chromosomes is also referred to as an organism's **genome**. The modern discipline of **genomics** represents the study of a species' genome, usually from a molecular perspective. Gametes are not, however, the only haploid cells. Many plants show an alternation of two distinct generations within their life cycle. The cells of the spore-producing **sporophyte** generation are diploid, while those of the gamete-producing **gametophyte** generation are haploid (Fig. 5.3).

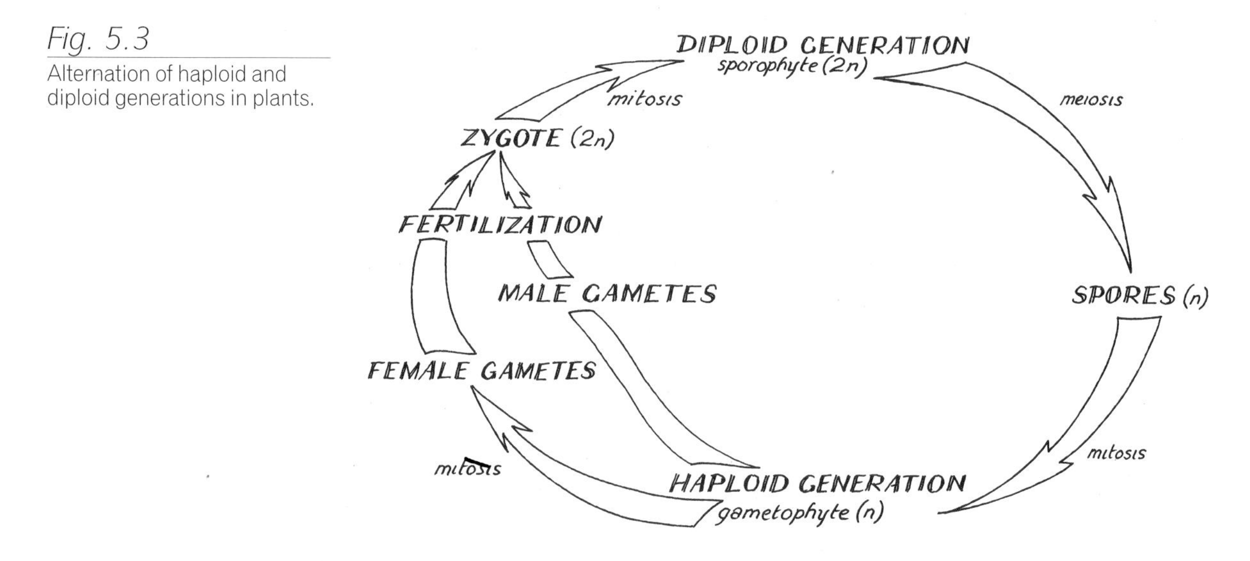

Fig. 5.3
Alternation of haploid and diploid generations in plants.

It is important to remember that, although each pair of chromosomes in a diploid cell possesses the same sequence of genes, the two chromosomes are not usually identical. This is because individuals are heterozygous at many loci, i.e. possess a different allele on each chromosome. Those rare examples, where virtually all loci are homozygous, produce intensely inbred populations such as the African cheetah or the Hawaiian goose, the nay-nay.

5.5 Two types of nuclear division

Because chromosomes carry hereditary information it is vital that, each time a cell divides during growth of an organism, both daughter cells receive a complete and uncorrupted copy of this vital information. A precisely controlled process of nuclear division called **mitosis** ensures that each daughter cell receives the same number and types of chromosomes as were present in the parental cell.

The life cycle of most organisms includes sexual reproduction. There must, therefore, be a stage in the life cycle when the chromosome number is halved, so that gametes contain a haploid set of chromosomes. If that did not occur, then the chromosome number of a species would double each generation, at fertilization. **Meiosis** is the nuclear division resulting in cells with half the chromosome number of the parental cells. Meiosis is thus essential for maintaining the balanced diploid chromosome number of a species. The next four sections consider the processes of mitosis and meiosis in more detail.

5.6 Mitosis

The basic features of mitosis are the same in all organisms. At the beginning of mitosis each chromosome is present as a replicated double structure, i.e. as two chromatids. A highly organized sequence of events follows that ensures that each pair of chromatids separates at the same time, and that one chromatid of each pair finds its way into each daughter nucleus.

Mitosis is a continuous process. However, when describing mitosis, it is more convenient to recognize four main stages – **prophase**, **metaphase**, **anaphase** and **telophase**. The key events of these four stages are illustrated in Fig. 5.4. As discussed earlier (Section 5.2), most nuclei contain many chromosomes. When describing mitosis the key features are more clearly conveyed by showing cells with a diploid chromosome number of four ($2n = 4$).

Prophase is the longest stage of mitosis. During this phase the long chromatin threads coil up, becoming shorter and thicker and recognizable as individual chromosomes. Other events are also occurring to make sure that there will be an equal and accurate distribution of chromosomes to the two daughter cells. Centrioles, a collection of tiny microtubules, divide and move to opposite ends or **poles** of the cell. Long fibres develop from the centrioles. Some of these **spindle fibres** extend from pole to pole, while others link up with the chromosomes and ensure a complete set of chromosomes segregates to each

Fig. 5.4

Mitosis in a generalized animal cell.

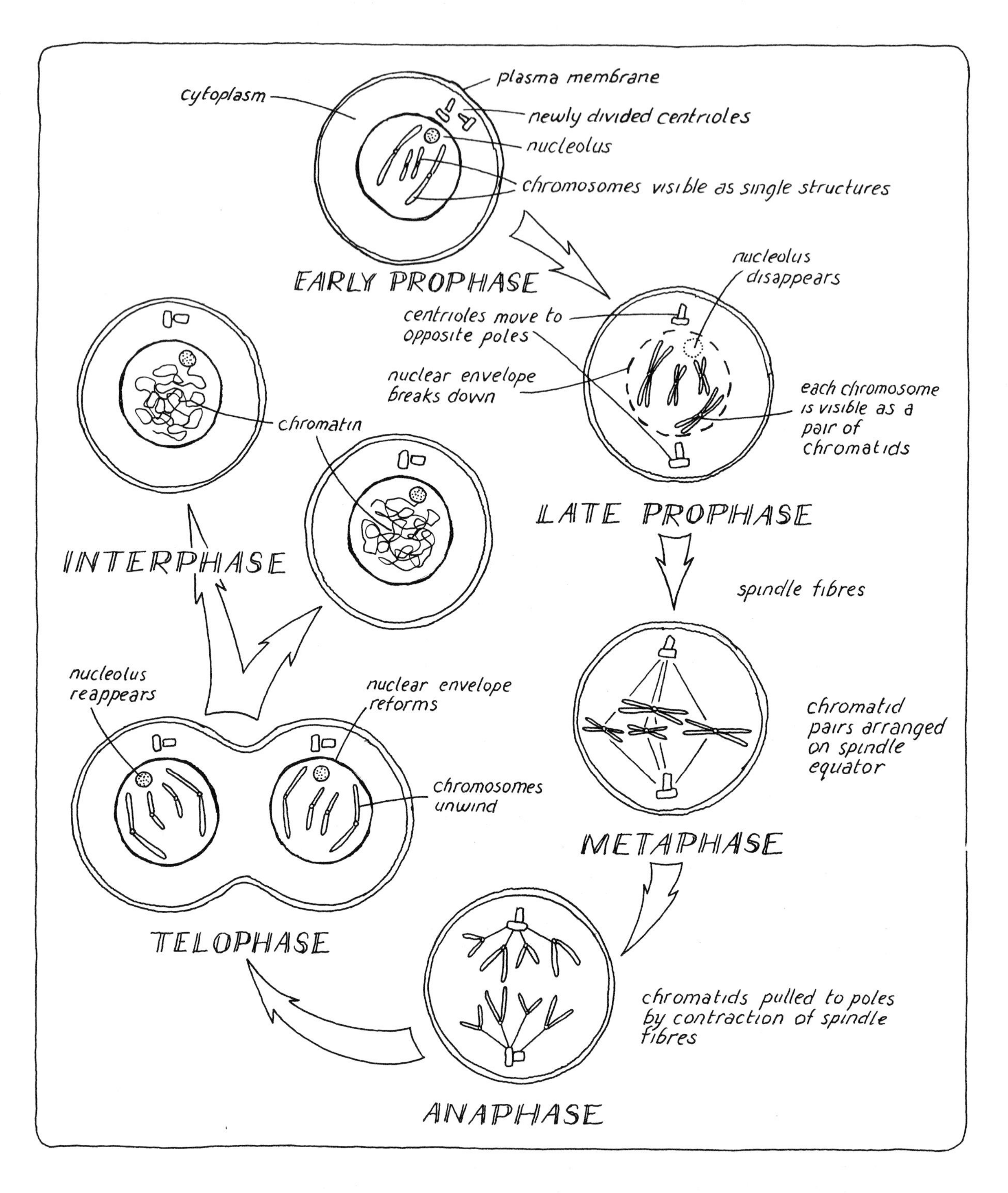

new nucleus*. A star of short fibres, forming the **aster**, can also be seen radiating from each centriole (see Fig. 5.5). Towards the end of prophase each chromosome is seen to consist of two chromatids: chromosome replication occurs before mitosis but the new chromosomes remain closely associated. Thus early in prophase each chromosome is only visible as a single structure (Fig. 5.4). As the individual chromatids become visible the nucleolus disappears and the nuclear envelope breaks up into a number of small vesicles.

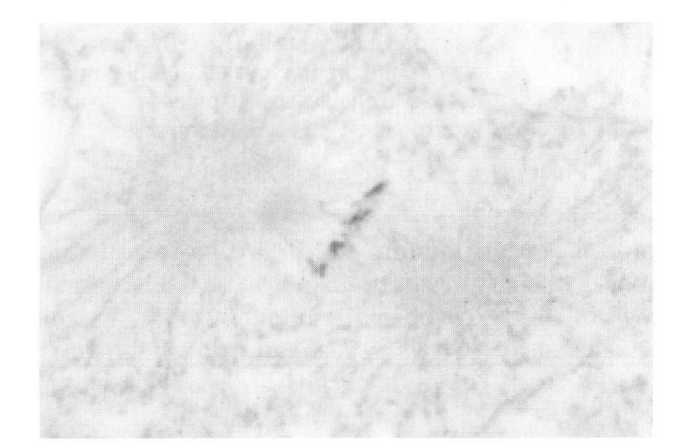

Fig. 5.5
Photomicrograph of metaphase in a whitefish cell. Spindles can be seen extending from the pole towards the chromatids on the equator. Radiating asters are also prominent.

The chromatid pairs now lie free in the cytoplasm. They gradually line up in the centre of the cell, or **equator**. At this stage, **metaphase**, the pairs of chromatids are held at the equator by the attachment of spindle fibres to their centromeres. **Anaphase** soon follows. The centromeres divide and the attached spindle fibres contract, pulling the separated chromatids to opposite ends of the cell. The reverse of prophase now happens. The chromosomes uncoil to once again form long diffuse chromatid threads. New nuclear envelopes form around the two separated groups of chromosomes among which a new nucleolus appears. This reorganization phase is known as **telophase**. The behaviour of the chromosomes during the different stages can be clearly seen in Fig. 5.6.

Mitosis is followed by **cytokinesis**, or cell division. In animal cells this simply involves constriction of the cytoplasm between the two new nuclei. Dividing plant cells have to construct a new wall. Thus, vesicles containing wall material gather in the middle of the cell and fuse, producing the **cell plate** against which new walls are constructed. The period between two nuclear divisions is known as **interphase**. The cycle of events, from one mitosis to the next, is referred to as the **cell cycle** (Box 5.2).

---

*Strangely, no centrioles have ever been observed in higher plant cells, yet they form a spindle apparatus.

their consequences are discussed in Section 7.3. Towards the end of prophase I, as in mitotic prophase, the nucleolus disappears, the nuclear membrane fragments and a spindle apparatus is constructed.

Homologous pairs of chromosomes remain associated at chiasmata. It is, therefore, pairs of chromosomes that align themselves on the equator of the spindle during **metaphase I**. This contrasts with mitotic metaphase when homologous chromosomes move independently of each other; pairs of chromatids arrange themselves on the mitotic equator. During **anaphase I** the homologous chromosomes separate and pairs of chromatids are pulled to opposite poles. Each new nucleus thus receives a haploid set of chromosomes, although each chromosome is present as a pair of chromatids (Fig. 5.9). The second meiotic division separates these pairs of chromatids.

Following **telophase I**, when a nuclear envelope reforms around each group of haploid chromosomes, cell division occurs. The resulting cells may proceed directly into **meiosis II** or there may be a rest phase, the length of which is species-specific. The events of meiosis II are essentially the same as those of mitosis (Fig. 5.4). Pairs of chromatids align on the equator and are pulled apart during anaphase II. However the cells that are produced at the end of the process are haploid: the number of chromosomes has been halved compared to those present at the beginning of meiosis (Fig. 5.10).

5.9 The biological significance of meiosis

Meiosis ensures constancy of chromosome number from generation to generation. Prior to fertilization it results in haploid cells so that when gametes fuse, a diploid zygote is produced. Meiosis also ensures variability among progeny because it promotes new combinations of alleles. This is a consequence of:

1. Independent assortment of chromosomes at metaphase I. This process varies which of two or more possible alleles at one locus is found with which allele at another locus on a different chromosome (discussed further in Section 5.10).
2. Swapping of pieces of homologous chromosomes during prophase I. This results in new combinations of alleles on the **same** chromosome (more on this in Chapter 7).
3. Random fertilization of gametes.

Together these three processes achieve maximal reshuffling of the available genetic material. It is, however, only mutations that produce new variants, or alleles.

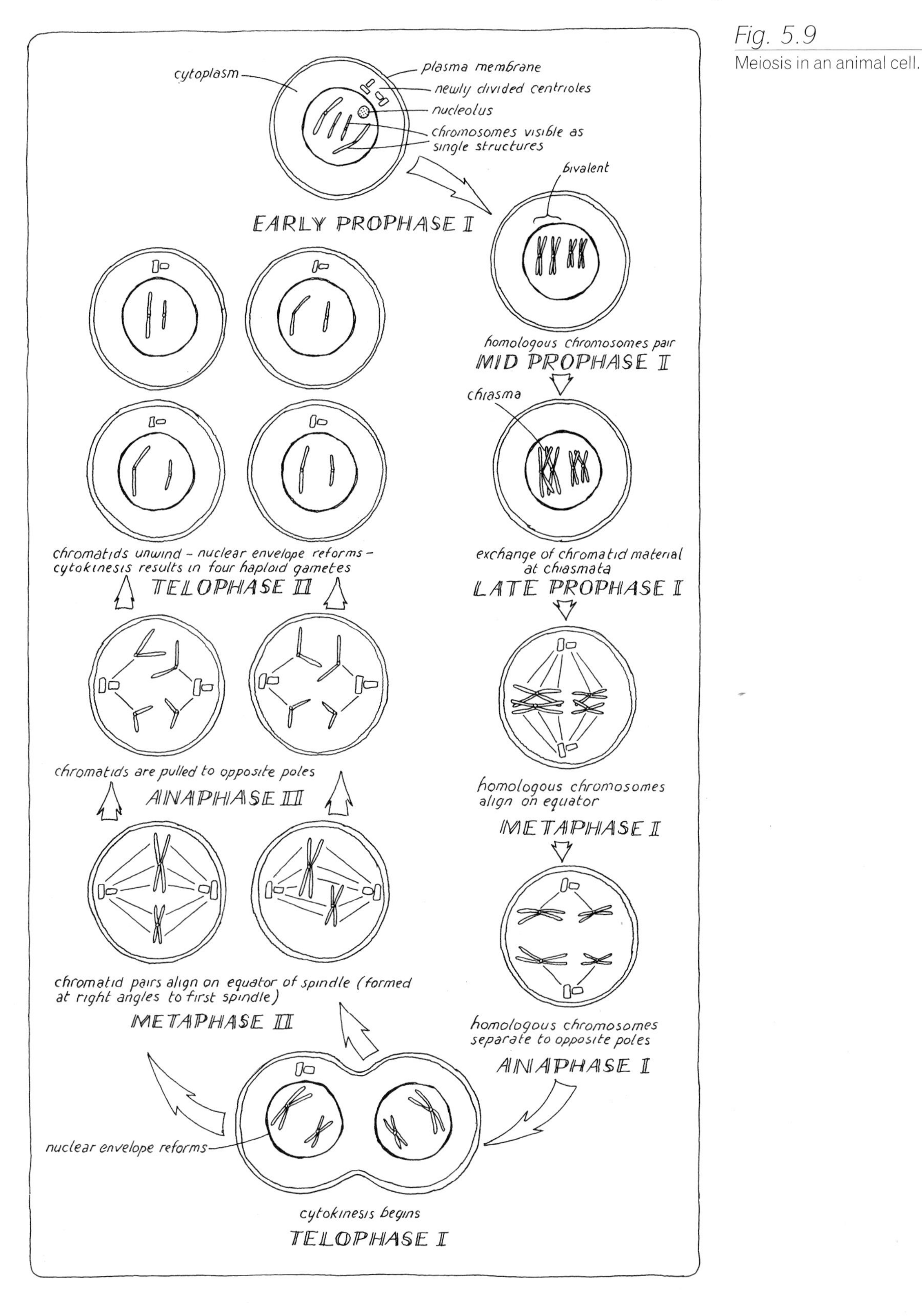

Fig. 5.9
Meiosis in an animal cell.

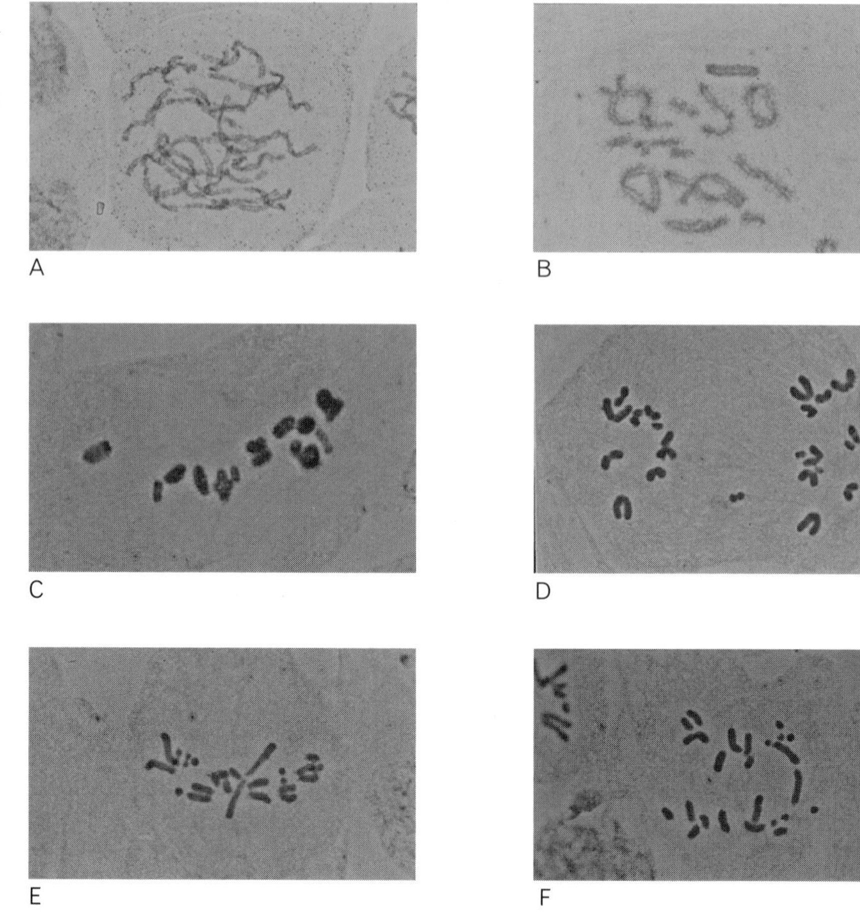

Fig. 5.10

Photomicrographs of meiosis in the testis of a grasshopper. A is early prophase I where the full complement of chromosomes, each consisting of a pair of chromatids, can be seen. B is later in prophase I where cross-overs have occurred between non-sister chromatids; C is metaphase I; D is anaphase I; E is metaphase II; and F is anaphase II.

5.10 Revisiting Mendel's laws

The key ideas of Mendelian inheritance – that pairs of alleles segregate into separate gametes, and that different pairs of alleles assort independently of each other – can be explained by the behaviour of chromosomes during meiosis.

Mendel's first law states that when an organism forms gametes, only one of a pair of alleles enters each gamete. The process of meiosis explains this segregation of alleles. Consider one gene with two alleles **A** and **a**. This gene will be found at a particular locus on a chromosome. Figure 5.11 illustrates key aspects of the behaviour of the relevant pair of homologous chromosomes, and therefore alleles **A** and **a**, during meiosis in a heterozygote.

We can next consider what happens when two heterozygotes of genotype **Aa** mate. As shown in Fig. 5.11, each heterozygote would produce the two possible types of gametes in equal proportions. Fertilization is a random event. Thus, four outcomes are equally likely, which, as Fig. 5.12 reminds us, results in two possible phenotypes expected to occur in a ratio of 3 : 1. This key pattern of monohybrid inheritance (Section 2.1) is thus explained by the behaviour of chromosomes during meiosis.

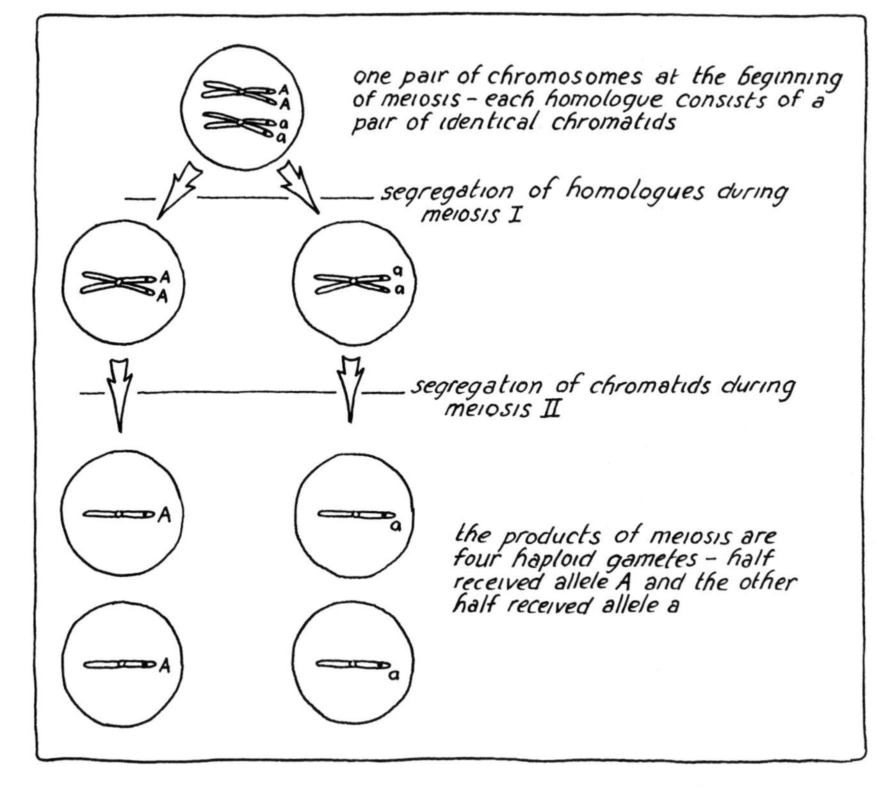

Fig. 5.11

Meiosis segregates pairs of alleles into separate gametes.

Heterozygote parents **Aa** **Aa**

Possible gametes **A** a **A** a

Gametes	A	a
A	**AA** dominant	**Aa** dominant
a	**Aa** dominant	**aa** recessive

Offspring phenotypic ratio: 3 dominant : 1 recessive

Fig. 5.12

Predicting the outcome of heterozygote matings.

From the results that Mendel gained from his dihybrid crosses he produced his Law of Independent Assortment, which states that either of a pair of alleles can segregate into a gamete with either of another pair. The key event in meiosis that explains this conclusion is the behaviour of chromosomes at metaphase I. Figure 5.13 shows that when two pairs of homologues align themselves on the equator, there are two different ways that the homologue pairs can orientate themselves relative to each other. As a consequence of these two possible arrangements of the homologues at metaphase I, four different types of gametes are possible. Considering a heterozygote of genotype **AaBb**, allele **A** could segregate into a gamete with either allele **B** or **b**, and, likewise, allele **a** could be found in a gamete with either allele **B** or **b**. These various assortments produce gametes of four different genotypes **AB**, **Ab**, **aB** and **ab**. However, Fig. 5.13 also shows that any one meiosis will only produce two types of gametes with respect to allele combinations at two loci on separate chromosomes. It is as the result of many meioses in a large group of heterozygotes that the four gametic types are produced in equal numbers.

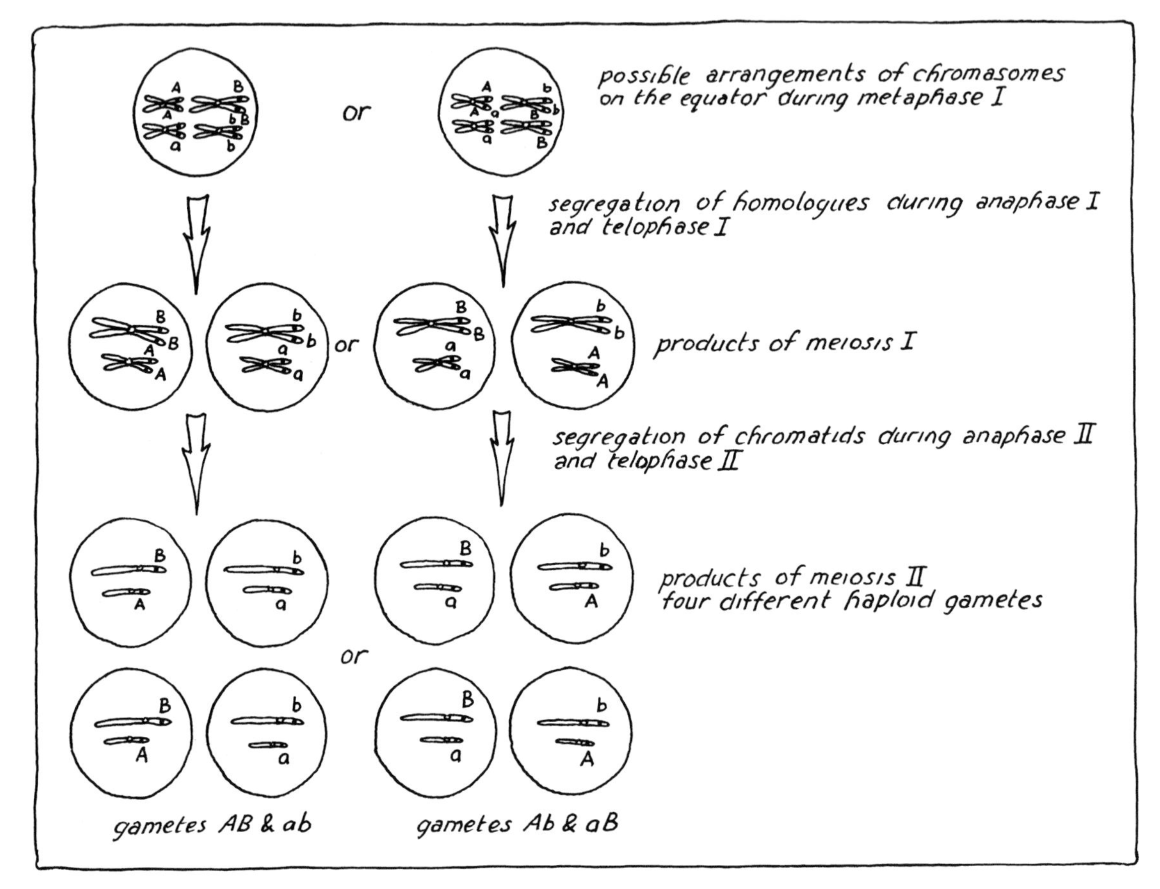

Fig. 5.13

The two possible outcomes of meiosis in a heterozygote of genotype AaBb.

Because, during sexual reproduction, any one of the four types of gametes produced by a female double heterozygote can be fertilized by any one of the four types of gametes produced by a male double heterozygote, there are 4×4, or 16, possible combinations of gametes. It may be remembered that these 16 possible fertilizations produce nine different genotypes, which fall into four different phenotypic classes with an expected frequency ratio of 9 : 3 : 3 : 1 (Fig. 4.2). Again, a key pattern of Mendelian inheritance is explained by the behaviour of chromosomes during meiosis.

Summary

- Genes are situated on chromosomes in the cell nucleus.
- Genes are found at specific loci on chromosomes.
- Chromosomes occur in homologous pairs in diploid cells and singly in haploid cells.
- The number of homologous pairs of chromosomes varies between species.
- The process of mitosis produces new nuclei with the same number and types of chromosomes as the parental one.
- The process of meiosis produces nuclei with half the number of chromosomes as the parental one.
- The behaviour of chromosomes during meiosis explains Mendel's Laws of Segregation and Independent Assortment.

Problems

1. Specify whether the following events occur during mitosis, meiosis I and/or meiosis II:

(a) pairing of homologous chromosomes

(b) alignment of chromosomes along the equator

(c) separation of sister chromatids

(d) attachment of pairs of sister chromatids to spindle fibres.

2. The mosquito *Culex pipiens* has a diploid chromosome number of six. Draw diagrams to show the arrangement of chromosomes during:

(a) anaphase II of meiosis

(b) metaphase of mitosis

(c) metaphase I of meiosis.

3. List six important differences between mitosis and meiosis.

4. (a) What are (i) homologues; (ii) sister chromatids?

(b) How similar to and different from each other are homologues and sister chromatids, respectively, with regard to genes and alleles?

5. Diploid cells of the fruit fly, *Drosophila melanogaster*, contain eight chromosomes. How many different random arrangements of homologues could occur during metaphase I of meiosis?

6. In the Madagascar periwinkle, *Catharanthus roseus*, there are six chromosomes per haploid set. How many chromosomes would you expect to observe if you prepared karyotypes of the following: (a) a leaf cell; (b) the sperm nucleus of a pollen grain; (c) a petal cell?

7. Match events from the list below with the appropriate stage of the cell cycle:

(a) Growth period between replication and nuclear and cell division; each chromosome has two sister chromatids.

(b) Growth period between division and replication; each chromosome has one chromatid.

(c) Chromosome replication.

(d) Nucleus undergoes mitosis followed by cell division.

Stages of the cell cycle: (i) G1; (ii) S phase; (iii) G2; (iv) M phase

8. Indicate which of the cells (2n = 6) in the figure corresponds to the following meiotic stages:

(a) prophase I; (b) metaphase II; (c) anaphase I.

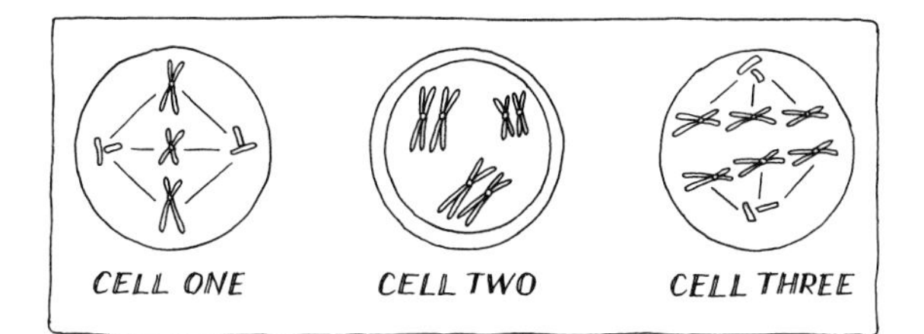

9. Chimpanzees have 48 chromosomes in each somatic cell.

(a) How many chromosomes does a baby chimpanzee receive from its mother?

(b) How many autosomes and how many sex chromosomes are present in each somatic cell?

(c) How many chromosomes are present in each sperm?

10. A Shetland pony, with a diploid set of 64 chromosomes, was sharing a field with a male zebra, which has 44 chromosomes in each diploid cell. Unexpectedly the Shetland pony gave birth to a foal. How many chromosomes would you expect in each somatic cell?

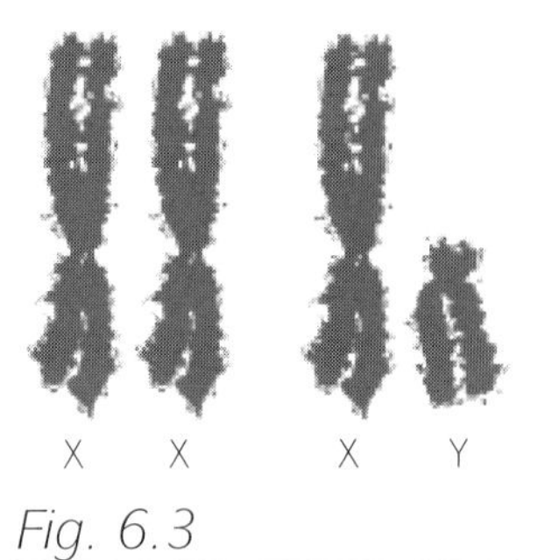

Fig. 6.3
Photomicrograph of human X and Y chromosomes.

There is, however, a small segment at the tip of the Y chromosome that is homologous with the X chromosome. This limited homology ensures a male's sex chromosomes pair during prophase I of meiosis and so segregate into separate cells during anaphase I. Thus, one product of the first meiotic division receives an X chromosome and the other a Y. Without this pairing mechanism for the X and Y chromosomes, gametes would be formed without, or with too many, sex chromosomes. As a consequence there would be problems with determination of the sex of embryos resulting from fertilization. As it is, 50% of sperm receive an X chromosome and 50% a Y; and we have an explanation for the (approximately) equal sex ratio among mammalian offspring (Fig. 6.4).

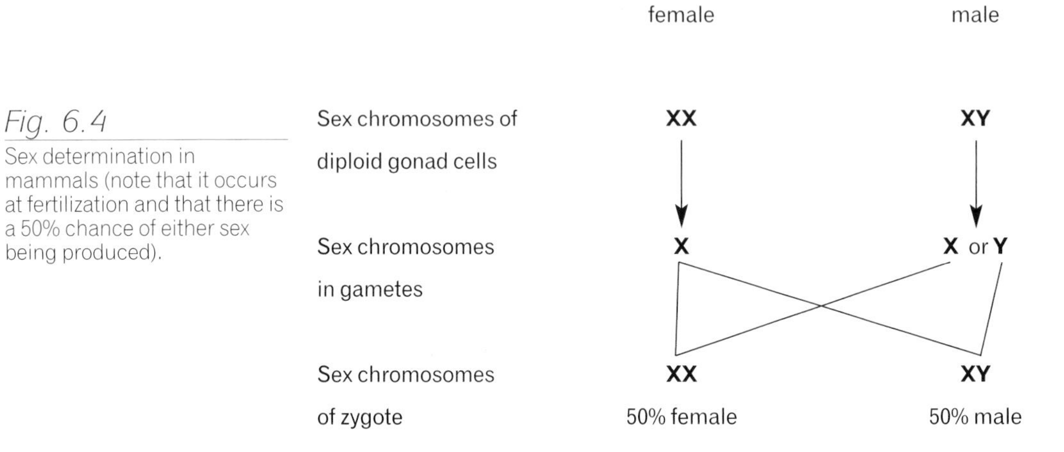

Fig. 6.4
Sex determination in mammals (note that it occurs at fertilization and that there is a 50% chance of either sex being produced).

Figure 6.4 shows that all the female gametes carry the same kind of sex chromosome. Female mammals are therefore sometimes described as the **homogametic sex**. Males produce two different types of gametes and are referred to as the **heterogametic sex**. By contrast, in birds, butterflies, moths and some reptiles and fish, it is the female who is the heterogametic sex and the male the homogametic one. To avoid confusion, the sex chromosomes in these groups of animals are called **W** and **Z**: the female is ZW and the male ZZ.

6.2 Other sex-determining mechanisms

The presence of a pair of similar or dissimilar sex chromosomes is not the only means by which the sex of an individual is determined. A variety of other sex-determining mechanisms exist throughout the living world. In some insect species, e.g. the grasshopper, there is only one type of sex chromosome, the X. Females are produced when two X chromosomes are present (**XX**). Just one X chromosome (**XO**) results in a male. Ants, bees and wasps lack any sex chromosomes. Instead, the sex of an individual is determined by its number of chromosome sets. Diploid individuals are female, while males are haploid (rare examples of haploid animals!).

In some lower invertebrates a single locus with two alleles determines gender. For example, in mosquitoes, males are heterozygous and females homozygous recessive for a sex-determining gene. Related to this are the mating type genes found in fungi. There are not separate male and female fungi, yet individual hyphal networks must be of different mating types to successfully fertilize. Mating type is determined by alleles at a single locus.

There are even a few examples, scattered around the animal and plant kingdoms, where the sex of a developing individual is determined by environmental factors. Reptile eggs exposed to high temperatures produce mostly males in some species and females in others (Fig. 6.5)!

Fig. 6.5

A female Morelets crocodile, *Crocodile morelets*, on her nest mound. The sex of her offspring will be determined by the temperature in the nest at a critical period.

So might the dinosaurs have become extinct because the temperature rose and only one sex was produced?! The sex of some fish is determined by social dominance whilst that of various marine worms and gastropods depends on the substrate upon which larvae settle. Certain plant species produce male or female flowers depending upon day length, although more usually sex determination of flowering plants has a chromosomal basis. About 90% of flowering plants are hermaphrodite (i.e. have both male and female flowers). Thus a sex-determining mechanism is irrelevant. Of the remaining 10% some species are **monoecious** – bearing separate male and female flowers on the same plant – while other species are **dioecious** – having separate male and female plants. The holly, *Ilex aquifolium*, is an example of a dioecious plant. If you have ever felt frustrated at Christmas by the tree in your garden that never bears any bright red berries, and so is useless for decorating the house, now you know the reason: it is because it is a male tree! Among monoecious and dioecious plant species sex-determining mechanisms vary, as they do with animals and, like animals, they seem to promote sexual development along one pathway or another. All flowers seem to have the potential to be hermaphrodite. Depending upon the sex chromosomes present, full development of either female or male flower parts is promoted and of the other repressed. So if an XY system is operating the Y chromosome carries genes that are needed for development of male flower parts while suppressing development of female parts.

6.3 The Y chromosome and sex determination in humans

Although the Y chromosome is much smaller, and carries many fewer functional genes than the X chromosome, it is the presence of a Y chromosome that is the trigger for development of the male phenotype. The crucial role of the Y chromosome was first suggested by the phenotypes of individuals with unusual sex chromosome karyotypes.

Occasionally segregation of the sex chromosomes does not occur properly during anaphase I or II of meiosis (more in Chapter 8). Gametes result that lack or contain extra sex chromosomes, and likewise the zygotes produced at fertilization. For example, if the sperm that fuses with the egg lacks a sex chromosome, the resulting zygote will be **XO**. About 5% of XO zygotes develop normally. The individual is female (Turner's syndrome). If a normal ovum (X) is fertilized by an XY sperm, the zygote will be **XXY** and develops into a male (Klinefelter's syndrome). These two karyotypes and associated phenotypes suggest that the presence of a Y chromosome is the switch needed for development along a male pathway because:

1. XO individuals have only one X chromosome, as in normal males (XY), yet they are female.
2. XXY individuals have two X chromosomes, as in normal females (XX), but are males.

Thus, the absence of a Y chromosome results in a female, while its presence promotes the development of a male. In 1991 the key male-determining gene on the Y chromosome was identified.

Very rarely, apparently 'sex-reversed' individuals are identified, i.e. **XX males** and **XY females**. In such cases one of the X chromosomes of an XX male invariably has an extra segment, derived from the Y chromosome. XY females generally lack the same region on their Y chromosome. In this critical region the crucial male-determining gene, the **SRY** gene, was discovered. During the early stages of embryological development the gonads are capable of developing into either ovaries or testes. The embryo's hormonal environment decides their fate. A functioning SRY gene triggers the production of testosterone and the gonads develop into testes. Femaleness is, therefore, the default pathway!

Sex chromosomes, therefore, determine the sex of an individual. When the XY system is operating, an active SRY gene on the Y chromosome is crucial in triggering development of a male phenotype. Sex chromosomes do not, however, carry all the genes responsible for sexual characteristics: they are scattered throughout the autosomes. Likewise, genes determining non-sexual characteristics are found on sex chromosomes, in particular on the X chromosome. Because females possess two copies of an X chromosome and males only one, this results in genes on the X chromosome having a different pattern of expression in the two sexes. The recessive phenotype is much more common in males.

6.4 The expression of X-linked genes

When considering **X-linked genes**, i.e. genes located on the X chromosome, the possible genotypes are different for males and females. Males cannot be homozygous or heterozygous for alleles of an X-linked gene as they can only possess one allele. Males are instead **hemizygous**. Figure 6.6 considers the possible genotypes produced by alleles of one of the many different genes found on the human X chromosome, i.e. of the gene that codes for production of clotting factor VIII. The gene possesses two main alleles: the dominant **H** allele produces normal clotting factor while the recessive allele **h** leads to a lack of this crucial protein. The blood of anyone who is homozygous or hemizygous for the recessive allele fails to clot. They suffer from the disease haemophilia*.

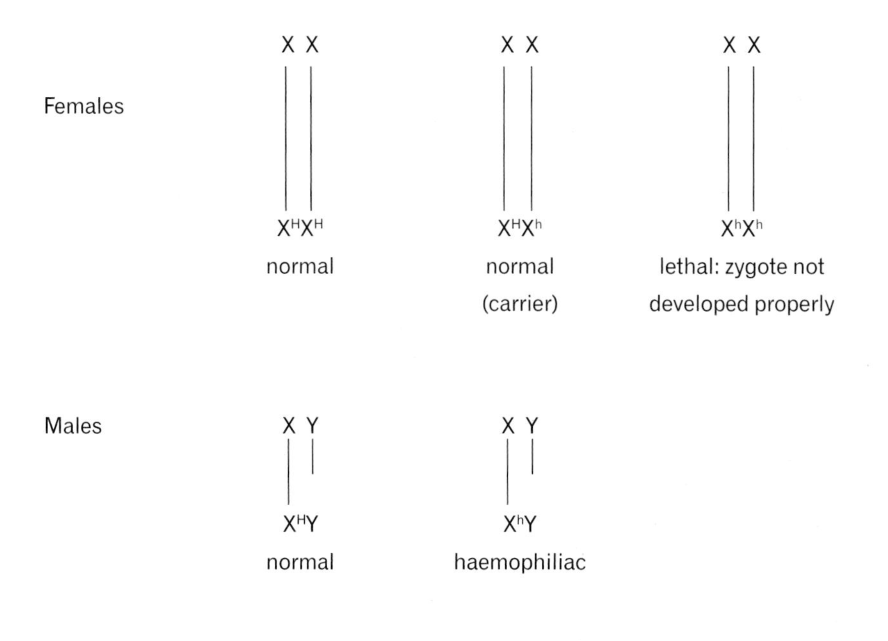

Fig. 6.6
The possible genotypes and associated phenotypes for haemophilia A in humans. When representing sex-linked genes it is best to show the allele symbols on an X chromosome, and to include the male Y chromosome. Both these procedures reduce errors when considering transmission patterns as they remind us that there is no copy of the relevant gene of the Y chromosome.

*There are several different inherited forms of haemophilia, each one resulting from a different defect in the biochemical pathway forming fibrinogen. The genes for two key proteins in this pathway are located on the X chromosome: clotting factor VIII and Christmas factor. When these factors are non-functional, the result is haemophilia A and haemophilia B respectively.

6.5 Representing crosses involving X-linked genes

We can use genetic diagrams to show how X-linked genes are inherited in the same way as they were used for investigating patterns of inheritance of genes on autosomes. Figure 6.7 shows how a haemophiliac child can be born to parents, neither of whom express the condition.

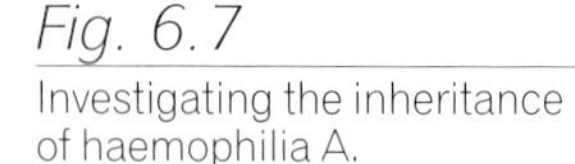

Fig. 6.7
Investigating the inheritance of haemophilia A.

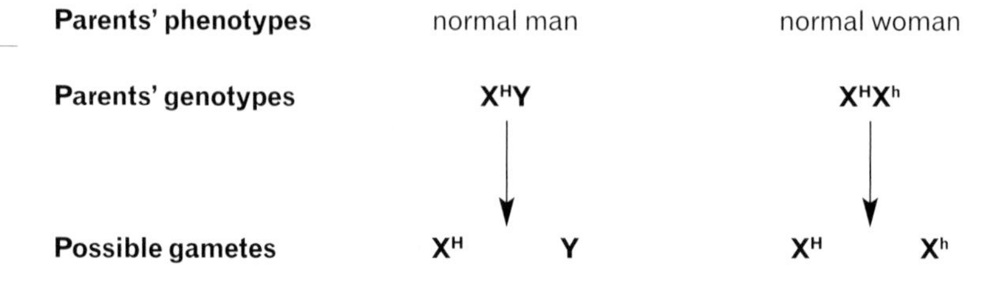

Possible genotypes and phenotypes among children

Gametes	X^H	X^h
X^H	X^HX^H normal	X^HX^h normal (carrier)
Y	X^HY normal	X^hY haemophiliac

Thus, if the woman is heterozygous, or a **carrier** for the recessive allele, then, as shown in Fig. 6.7, each time the couple conceive a child they have a 1 in 4 chance of their child being a haemophiliac. Expressed another way, if the couple have a daughter her blood will definitely clot normally, but if they have a son he has a 1 in 2, or 50%, chance of being a haemophiliac. This example illustrates the greater expression of X-linked conditions in males; a phenomenon discussed further in the next two sections.

6.6 Sex-linked inheritance patterns

In the examples of monohybrid and dihybrid inheritance discussed in Chapters 2 to 4, reciprocal crosses gave the same results. For example, when the inheritance patterns shown by a pair of contrasting characters (e.g. red- and yellow-fruiting tomatoes) were being investigated it did not matter which parent showed which phenotype. By contrast, when sex-linked characters are considered, reciprocal crosses give different results. Consider, for example, reciprocal crosses investigating the inheritance of eye colour in the fruit fly *Drosophila melanogaster* (Fig. 6.8). Normally the eyes of this fly are brick-red. White eyes sometimes occur because of a mutation in a gene that codes for a transmembrane transporter protein. The changed protein no longer transports the red pigment precursors into the eye pigment cells, so eyes appear white.

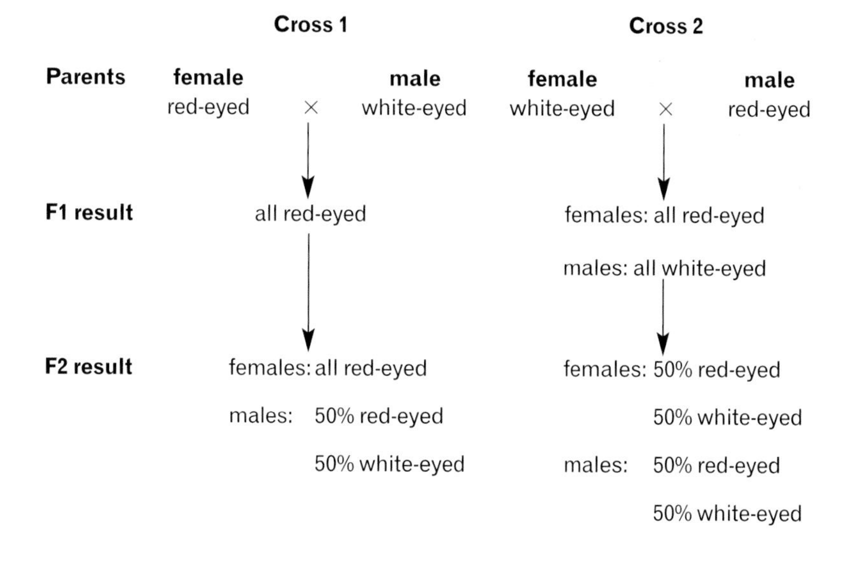

Fig. 6.8
Investigating the inheritance of eye colour in *Drosophila melanogaster*: the results of reciprocal crosses.

The F_1 result of cross 1 (Fig. 6.8) clearly shows that red eyes are dominant to white eyes. The fact that the F_1 and F_2 results are different depending upon which parent was white-eyed and which red-eyed indicates that the encoding gene is sex-linked. Figure 6.9 shows the genotypes of the flies, thus illustrating some other features associated with sex-linked genes, namely that:

- Heterozygous females transmit their X-linked recessive allele to approximately half of their daughters and half of their sons. Expression only occurs in their sons (see crosses 1 and 2).
- Males that inherit an X-linked recessive allele exhibit that trait as their Y chromosome has no counterpart. In contrast females need two copies of a recessive allele to express the recessive trait (cross 2).

	Cross 1		Cross 2	
Parents	red-eyed female **X^RX^R**	white-eyed male **X^rY**	white-eyed female **X^rX^r**	red-eyed male **X^RY**
F_1 result	**X^RX^r**	**X^RY**	**X^RX^r**	**X^rY**
	all red-eyed		red-eyed females white-eyed males	
F_2 result	females: 50% **X^RX^R** (red-eyed) 50% **X^RX^r** (red-eyed)		females: 50% **X^RX^r** (red-eyed) 50% **X^rX^r** (white-eyed)	
	males: 50% **X^RY** (red-eyed) 50% **X^rY** (white-eyed)		males: 50% **X^RY** (red-eyed) 50% **X^rY** (white-eyed)	

Fig. 6.9
Investigating the inheritance of eye colour in *Drosophila*: genotypes of flies in Fig. 6.8.

Problems

1. If a woman heterozygous for a recessive sex-linked allele marries a man who does not show the trait what is the probability that:
   (a) the recessive allele will be passed on to any child the couple might have?
   (b) the couple might have a child who shows the trait?
   (c) any of their sons would show the trait?
2. What would be the most likely genotypes of parents whose daughters were all normal-sighted and whose sons were all colour blind?
3. A woman whose father suffered from haemophilia A is married to a normal man who thinks that his grandfather might have had the condition. What are the couple's chances of having affected children if the man's grandfather did in fact suffer from haemophilia?
4. Black fur in cats is a result of an X-linked allele, **B**, while the alternative allele, **Y**, results in ginger fur. Heterozygotes have a patchy black and ginger coat, referred to as tortoiseshell. What kittens might result from a mating of a black male and tortoiseshell female?
5. What advice might a genetic counsellor offer to a woman whose mother's brother died, aged 19, of the X-linked recessive condition, Duchenne muscular dystrophy? Her husband is in full health and there is no history of the condition in his family. She is concerned about the possibility of any children they might have suffering from the disease.
6. Among a litter of four Dalmatian puppies what is the probability that
   (a) two are males and two are females?
   (b) all four are females?
   (c) at least one puppy is a female?
7. Females have been observed with four Barr bodies in their cells. How many X chromosomes are present in each somatic cell?
8. If a rare genetic disease in horses is inherited on the basis of an X-linked dominant allele would an affected mare or an affected stallion always have affected female foals?

When answering Questions 9 and 10 remember that the female is the heterogametic sex.

9. Several pairs of grey cockatiels were crossed. In all cases the nests contained both white- and grey-feathered chicks, but all the white birds were female. When white-feathered birds were crossed all the resulting chicks were white. What is the probable basis of inheritance of feather colour in these birds?
10. It is generally impossible to sex chicks morphologically. However, commercial chicken breeders have devised a way, in some breeds, based on plumage colour. Barred plumage (**B**) is dominant over non-barred (**b**) or solid colour. This difference is encoded by a gene located on the Z chromosome. What must be the genotypes of the male and female parents in order to be able to sex the chickens soon after hatching?

7 Linkage and chromosome mapping

It is estimated that the human genome possesses in excess of 30,000 genes. These are distributed between just 23 different chromosomes. Therefore, a thousand genes, or more, must lie on each chromosome. A major goal of genetics at the beginning of the twenty-first century is to **map**, i.e. determine, the precise chromosomal position of all the human genes, as well as those of many other species. Mapping genes has become a sophisticated molecular process. Traditional Mendelian analysis has been, and continues to be, a useful tool in the initial stages, in that the results of crosses are used to produce preliminary or **outline** maps. These show which of an organism's genes are gathered together on a particular chromosome and their approximate positions relative to each other. Molecular analysis is then used to precisely locate these genes.

This chapter describes the contribution made to chromosomal mapping by Mendelian genetics, in particular how:

- distinctive dihybrid phenotypic ratios can indicate when two genes are located on the same chromosome;
- these ratios can be used to map genes relative to each other;
- the events of prophase I of meiosis results in the production of these ratios.

7.1 The discovery of linkage

During the 1900s William Bateson and Reginald Punnett performed a series of dihybrid breeding experiments with sweet peas aimed at investigating the inheritance patterns of a number of different pairs of characteristics*. They used the basic Mendelian technique of selfing pure-breeding lines to get an F_1 generation that they then interbred to produce F_2 plants. Some of these crosses yielded F_2 phenotypes in the expected 9 : 3 : 3 : 1 dihybrid proportions. In others the F_2 phenotypic ratios deviated strikingly from the proportions expected in a straightforward dihybrid situation. Furthermore no consistent ratio was found between different crosses. Table 7.1 shows the F_2 result from one such cross investigating the inheritance of flower colour and pollen grain shape. Flower colour was purple (**P**) or red (**p**); and pollen shape was long (**L**) or round (**l**). The pure-breeding parental plants bore purple flowers and long pollen grains or red flowers and round grains.

Table 7.1

F_2 phenotypes observed by Bateson and Punnett

Phenotype (and genotype)	Observed progeny	Expected progeny (if 9 : 3 : 3 : 1 ratio)
Purple long (**P-L-**)	4831	3911
Purple round (**P-ll**)	390	1303
Red long (**ppL-**)	393	1303
Red round (**ppll**)	1338	435

It can be seen that all four possible F_2 phenotypes are present, but two occur more frequently than expected. It is the combination of traits present in the parental generation that are prominent among the F_2 plants (i.e. plants with purple flowers and long grains, and plants with red flowers and round grains); while the other two phenotypic classes are under-represented (see Table 7.1). So how do we interpret Bateson and Punnett's result?

*You may remember that in another series of crosses, investigating the inheritance of white and purple sweet pea flower colour, Bateson and Punnett had shown how the action of genes can be complementary (see Section 4.5).

Figure 7.1 indicates that certain fertilizations among the F_1 gametes were more common than others (the unshaded part of the Punnett Square), or, put another way, the F_1 heterozygous plants produced predominantly **PL** and **pl** gametes (Figs 7.1 and 7.2).

Parental cross

purple flowers
long grain

PPLL

red flowers
round grain

ppll

F_1 result

PpLl × self

purple flowers
long grains

F_2 result

Gametes	P L	p l	P l	p L
P L	P L / P L purple long	P L / p l purple long	P L / P l purple long	P L / p L purple long
p l	P L / p l purple long	p l / p l red round	P l / p l purple round	p L / p l red long
P l	P L / P l purple long	P l / p l purple round	P l / P l purple round	P l / p L purple long
p L	P L / p L purple long	p L / p l red long	P l / p L purple long	p L / p L red long

Fig. 7.1
Bateson and Punnett's sweet pea cross (unshaded boxes represent the majority of plants observed).

F_1 heterozygote

Gametes produced

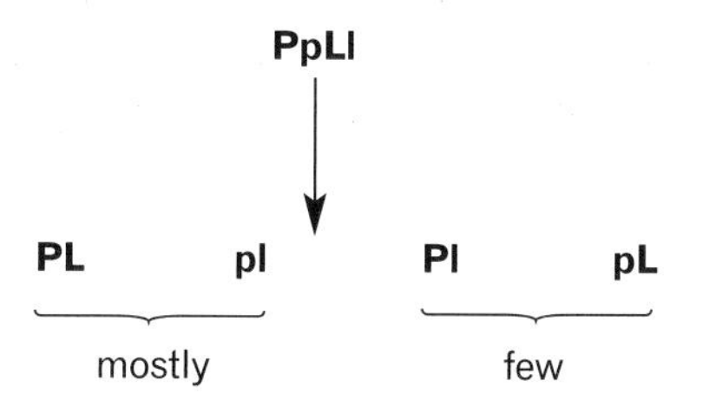

Fig. 7.2
Gametes produced by F_1 heterozygote.

Bateson and Punnett accounted for the greater production of two of the four possible types of gametes by the F_1 plants by suggesting that the dominant alleles (**P** and **L**) and the recessive alleles (**p** and **l**) were **physically coupled**, so that most of the time they had to be inherited together.

Without realizing it, Bateson and Punnett had chanced upon the genetic phenomenon of **linkage**, when two traits are inherited together because the loci of the genes controlling them are situated on the same chromosome. Indeed, as stated at the beginning of the chapter, gene linkage must be a widespread phenomenon, as in all species the number of genes far outstrips the number of chromosomes. The only logical conclusion is that many genes are gathered together on one chromosome. These interpretations came after Bateson and Punnett's time; they had no idea as to the nature of their proposed coupling. Indeed it was some years before their findings (that traits are sometimes inherited together) were even confirmed. This confirmation was largely due to the pioneering work of the geneticist Thomas Morgan (1866–1945). Working with the fruit fly *Drosophila melanogaster*, he also found major deviations from expected phenotypic ratios in dihybrid crosses. Morgan first proposed the idea of linkage and correlated the results of crosses involving linked genes with the behaviour of chromosomes during meiosis (Box 7.1).

Box 7.1 Thomas Morgan and the fruit fly

Thomas Hunt Morgan

Drosophila melanogaster

Thomas Hunt Morgan, an American, was one of the key geneticists of the first half of the twentieth century. Inspired by the rediscovery of Mendel's work, he began his own breeding experiments with the fruit fly, *Drosophila melanogaster.* He soon realized he had chanced upon an ideal organism for heredity work. It was extremely prolific, had a short generation time (10–14 days) and possessed many easily identifiable morphological variants; for example, different eye colours and shapes. Morgan's first breeding experiment, in 1910, was between a normal red-eyed fly and a newly discovered white-eyed mutant. The results convinced Morgan that genes are carried on chromosomes. He and his students, in particular Calvin Bridges, Hermann Muller and Arthur Sturtevant, followed this initial discovery with a decade of crosses that developed the methods for mapping genes on chromosomes and firmly established the fruit fly as a model organism for genetic study. Over the decades, the fruit fly has contributed enormously to many aspects of developmental and molecular biology. Its genome has been sequenced. A recent transgenic fly is, for example, providing valuable insights into the pathology of the complex human disease, Parkinson's disease. In 1933 Morgan was awarded the Nobel Prize for Physiology or Medicine.

7.2 Representing dihybrid crosses involving linked genes

Hitherto genotypes have been represented solely by letters; for example the double heterozygote as **AaBb**. When considering crosses involving linked genes it is generally better to indicate the genotype using a summary chromosome. This serves to remind us that we are dealing with linked genes. Thus, if genes **A** and **B** are located on the same chromosome, the double heterozygote would be represented as shown in Fig. 7.3.

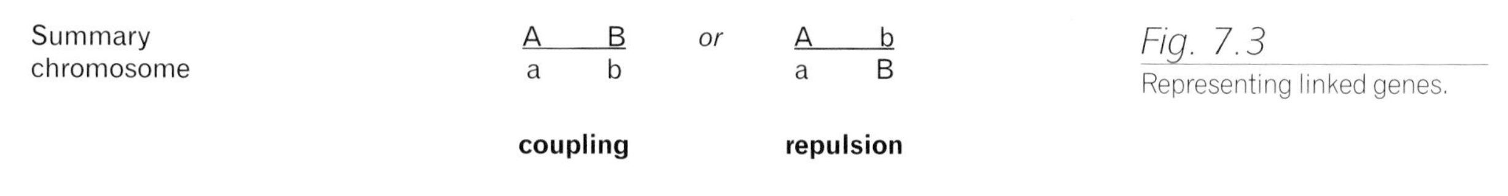

Fig. 7.3 Representing linked genes.

Figure 7.3 shows that two arrangements of the alleles are possible for a double heterozygote. When **coupled**, the two dominant alleles are linked on one chromosome and the two recessive alleles are on the other. **Repulsion** indicates a dominant and recessive allele on each chromosome. Using 'linkage format', Bateson and Punnett's original sweet pea cross can thus be represented as shown in Fig. 7.4 (the alleles are coupled!).

Gametes	**P L**	**p l**	**P l**	**p L**
P L	P L / P L purple long	P L / p l purple long	P L / P l purple long	P L / p L purple long
p l	P L / p l purple long	p l / p l red round	P l / p l purple round	p L / p l red long
P l	P L / P l purple long	P l / p l purple round	P l / P l purple round	P l / p L purple long
p L	P L / p L purple long	p L / p l red long	P l / p L purple long	p L / p L red long

Fig. 7.4 Representing Bateson and Punnett's sweet pea cross in linkage format.

7.3 Explaining Bateson and Punnett's sweet pea results

The majority, 89%, of Bateson and Punnett's F_2 sweet pea plants showed phenotypes within the unshaded area of the Punnett Square. This correlates with most of the gametes being parental types(**P L** and **p l**), and only

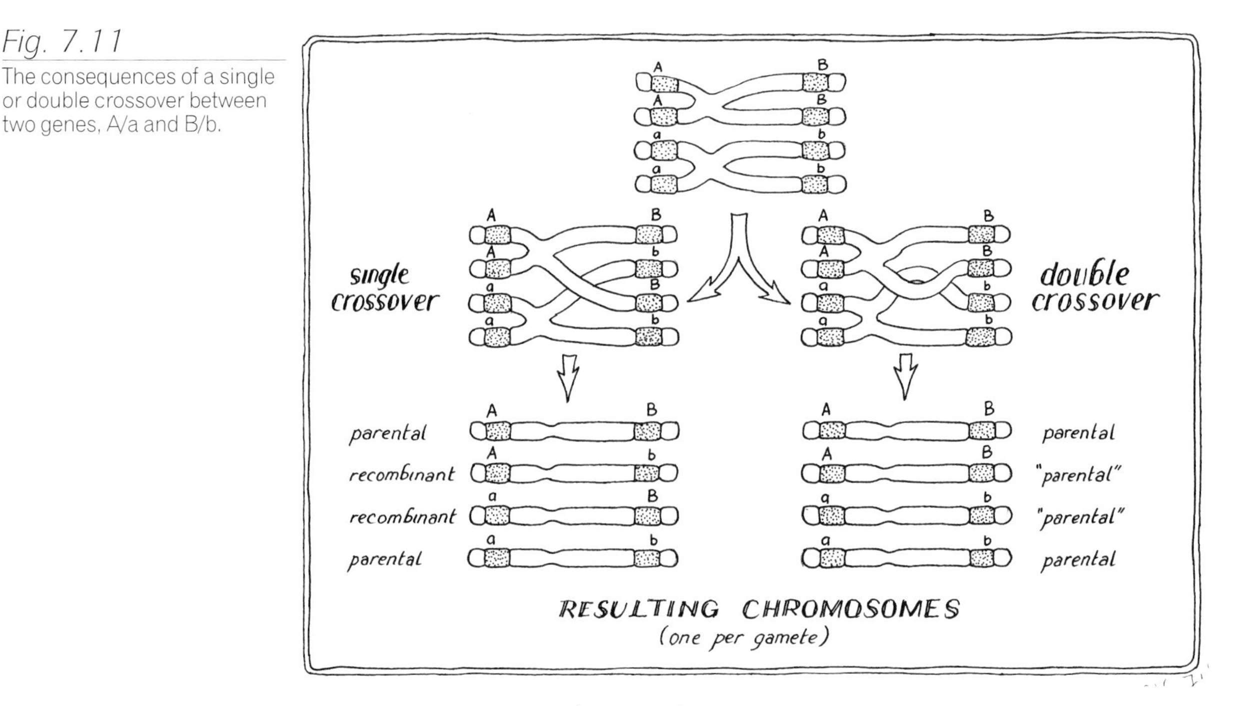

Fig. 7.11
The consequences of a single or double crossover between two genes, A/a and B/b.

a triple heterozygote (**AaBbCc**) and the recessive tester **aabbcc**. While a double crossover between genes A and C still returns the first and third pair of alleles to their original linkage position, the middle pair are exchanged (see Fig. 7.12). A rare class of recombinants is found among the offspring to be used in map distance calculations.

It is not the purpose of this book to go into further detail here. It is sufficient to say that we analyse the results of a three-point cross in a similar way to a two-point one. We consider each of the three different gene pairs in turn (**A** and **B**, **A** and **C**, **B** and **C**), and use the number of recombinants between each pair to work out map distances. It can actually be more efficient to perform three-point crosses: one such cross can yield the same result as three separate two-point crosses.

Not even a three-point cross can, however, eliminate a few other problematic crossover effects. Crossovers do not occur with the same frequency at all regions along a chromosome. There is a reduced chance of a crossover occurring near a centromere, at the ends of a chromosome and in regions where there are few functional genes (**heterochromatin**). This leads to an underestimation of distance between genes in these regions. In many species crossovers occur less often in males than females. Absolute distances between genes on male chromosomes are therefore less; relative distances will be the same – which, after all, is the relevant aspect of mapping. Chromosome maps are an abstract concept. They merely show the relative positions of genes. Other, **physical mapping**, techniques are then used to precisely locate genes on chromosomes.

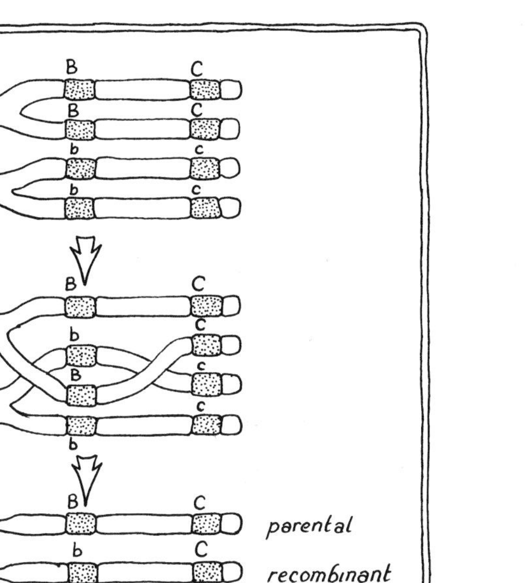

Fig. 7.12
The consequences of a double crossover in a three-point cross.

7.6 A few comments on physical mapping

As the twentieth century progressed, ever more detailed chromosome maps were produced for an increasing number of species. It is, however, only in the last couple of decades that geneticists have been able to start identifying the precise location of genes on their respective chromosomes. The DNA technological revolution has made this possible.

Molecular analysis of chromosomes has shown many sites of neutral variation at the DNA level, i.e. variants with no effect on an individual's phenotype, but detectable by molecular means. These sites are generally within non-coding regions of chromosomes and can be used as **DNA markers**. Individuals can be heterozygous for these marker regions, which we can refer to as 'alleles' and which are inherited in a Mendelian fashion. Crosses are made in which one of the 'genes' is a DNA marker whose precise location in a genome is already known. Thus mapping a gene, with known phenotypic effect, relative to a marker means we immediately have a definite idea of the gene's position. We can then:

1. Chop out the relevant segment of DNA.
2. Clone (i.e. make multiple copies) of this targeted segment for ease of analysis (see Chapter 13).
3. Determine the DNA sequence of the cloned segment and from its molecular character precisely locate our gene of interest.

Once the DNA sequence of a gene is known, it is possible to make predictions about the nature of the encoded protein and even its role within the cell. It is the gateway to therapies of various kinds – pharmacological and genetic.

Summary

- Many genes are located together on one chromosome: such genes are said to be linked.
- Dihybrid crosses do not always produce the expected 9 : 3 : 3 : 1 Mendelian ratio among the F_2 progeny.
- Deviations from the expected 9 : 3 : 3 : 1 F_2 ratio occur when the genes are linked.
- A particular sequence of alleles on a given chromosome is inherited together unless separated by crossing over during prophase I of meiosis.
- A high number of recombinants indicate widely separated genes.
- Crossing over produces recombinant gametes which, when used in fertilization, result in recombinant offspring.
- The number of recombinant offspring relative to parental types is a reflection of the frequency with which crossing over occurred between a pair of genes.
- Data collected from two-point test crosses for pairs of genes known to be on the same chromosome is used to produce a gene map showing their relative positions.
- Physical mapping can then precisely determine the genes' locations.

Problems

1. In corn snakes two genes on the same chromosome control colour. Any snake having the genotype **O-B-** is brown; **O-bb** is orange; **ooB**- is black and **oobb** is albino. A cross between an albino and a heterozygous brown snake (**OoBb**) produced 73 brown, 24 orange, 26 black and 69 albino offspring. How does this data indicate linkage between the two loci?

2. In the garden pea a gene controlling sensitivity to pea mosaic virus is linked to a gene controlling pod colour. Sensitivity to the virus is recessive to resistance and orange pods are recessive to green ones.
   (a) If plants true breeding for resistance and normal pods are crossed to others that produce orange pods and are sensitive to the virus, what will be the phenotype and genotype of the F_1 progeny?
   (b) If these F_1 progeny are test crossed what are the possible phenotypes of the progeny? As the genes are linked, which phenotypes will be the most frequent?

3. A recessive allele at a locus on chromosome 5 produces an odd gait in mice referred to as waltzing. When pure-breeding albino waltzer mice were mated with pure-breeding coloured non-waltzer mice, all the progeny had coloured fur and a normal gait. However, when these F_1 mice were crossed to a waltzer albino the following mice were obtained:

Phenotype	Number
Coloured, non-waltzer	36
Coloured waltzer	7
Albino non-waltzer	12
Albino waltzer	38

   (a) What evidence is there that the genes controlling these two traits are on the same chromosome?
   (b) State the genotypes producing each of the four phenotypes in the table.
   (c) Suppose an albino waltzer mouse was crossed with one of the albino non-waltzer mice – what progeny might you obtain?

4. Two loci are very close together on a chromosome. Suppose an individual recessive at both loci (**aabb**) was mated with a double heterozygote (**AaBb**). What phenotypes and in what proportions would you expect among the progeny if no crossing over occurred between these two loci during the meioses giving rise to the gametes used in this mating?

5. How is it possible for two genes on the same chromosome to undergo independent assortment?

6. If 7.7% of all meioses in winter wheat result in a chromatid exchange between loci C and D, what is the map distance in centimorgans between these two genes?

7. Green is the dominant colour in Pacific tree frogs. Occasionally a rare blue frog is found. This colour locus is linked to another determining the size of the frogs' toe pads. A large toe pad is recessive to a normal-sized one. In a cross between a blue, large-padded tree frog and a green individual with normal-sized pads, that is heterozygous at both loci, the offspring consisted of 32 green normal-padded frogs, 27 blue large-padded frogs, 9 green large-padded frogs and 12 blue normal-padded frogs. What is the map distance between the two loci?

8. In the fruit fly, *Drosophila melanogaster*, alleles at one locus determine the production of normal or modified antennae (aristapedia). Another pair of alleles results in grey or ebony bodies. Flies with normal antenna and grey bodies were crossed with others that had aristapedia antenna and ebony bodies. All the flies in the resulting F_1 generation had normal antennae and grey bodies. When F_1 flies were mated with the parental flies with aristapedia antennae and ebony bodies, approximately half the progeny had aristapedia antennae and ebony bodies and the other half normal antennae and grey bodies. In addition, in a few crosses, there were both flies with aristapedia antennae and grey bodies and normal antennae and ebony bodies. Explain these results.

9. Two genes are on the same chromosome, 14 cM apart. Parents of genotypes **AABB** and **aabb** were crossed and the resulting heterozygotes (**AaBb**) mated with recessive individuals (**aabb**).
   (a) Of the 2000 offspring produced, what numbers might you expect of the four different possible genotypes – **AaBb**, **aaBb**, **Aabb** and **aabb**?
   (b) What would be the predicted results of the test cross if the original parents had been **AAbb** and **aaBB**?

10. Three different genes were found to affect growth of mice tails. When either a fused tail allele (**F**) was present at one of these loci or a kinky tail allele (**K**) was present at a second, the resulting tails had a kinky appearance. Brachyury (**B**) mice have short tails. Crosses between individuals expressing the different alleles showed that these three genes were linked, close to the histocompatibility H_2 gene (**H**). The table shows the distances calculated as the result of a series of pairwise test crosses. Make a map of these four genes showing their relative positions to each other.

Genes involved in test cross	Distance between genes
F/K	1 cM
F/H	4 cM
B/H	12 cM
F/B	8 cM
K/H	3 cM

8 Variation in chromosomal number and structure

The genetic information of a diploid organism is delicately balanced in both content and location. Different genes are found at specific sites on specific chromosomes. Two copies are generally present at corresponding sites on the two homologues. The various alleles of a gene correlate with different phenotypes, and some are less efficient than others, occasionally even lethal. It is hardly surprising, therefore, that major changes to a chromosome, for example rearranging the genetic layout of a region, or even changing the number of chromosomes, can have a dramatic impact on an organism's phenotype, to the extent of preventing its normal development. Any change to the number or arrangement of chromosomes is known as a **chromosome mutation** or **chromosome aberration**.

This chapter considers chromosome aberrations from two perspectives:

- how they arise;
- their genetic and phenotypic consequences.

8.1 Changes in chromosome number: terminology

When an organism or cell has one or more complete sets of chromosomes it is said to be **euploid**. Thus, eukaryotic organisms such as mice and humans, which are normally diploid, can also be referred to as euploid. Likewise, any **polyploid** species with multiple chromosome sets is a euploid. Mutations can occur that reduce or increase the number of chromosomes of a set. Any organism or cell with a chromosome number that is *not* an exact multiple of the haploid number of chromosomes is described as **aneuploid** (Fig. 8.1).

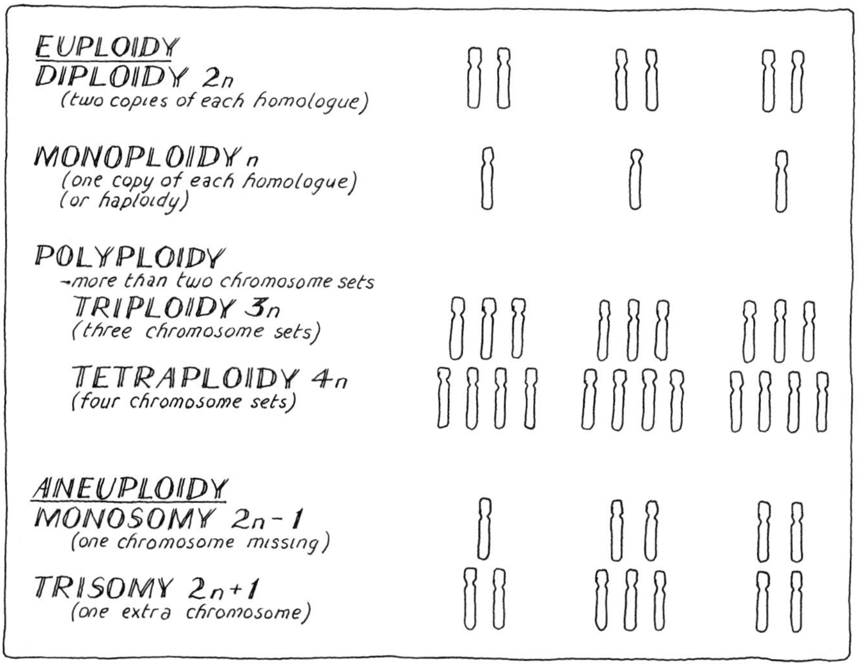

Fig. 8.1
Terminology associated with variation in chromosome number, illustrated by considering three homologous chromosome pairs.

8.2 Aneuploidy

The most frequent examples of aneuploidy are cases when a single chromosome is either lost from or added to a normal diploid set, i.e. **monosomy** or **trisomy**. The extra or lacking chromosome can be an autosome or a sex chromosome.

Aneuploidy usually results from **non-disjunction** during meiosis, when either a pair of chromosomes fail to segregate during anaphase I or a pair of chromatids do not separate during anaphase II. Figure 8.2 illustrates this process. Normally a gamete should have one copy of every chromosome. If non-disjunction occurs during the formation of a gamete it becomes **unbalanced**: lacking or containing two copies of a particular chromosome. Following fertilization of such gametes by a normal haploid gamete, monosomic and trisomic zygotes result (Fig. 8.2).

Fig. 8.2

Gametes and zygotes produced as the result of non-disjunction during the first or second meiotic divisions.

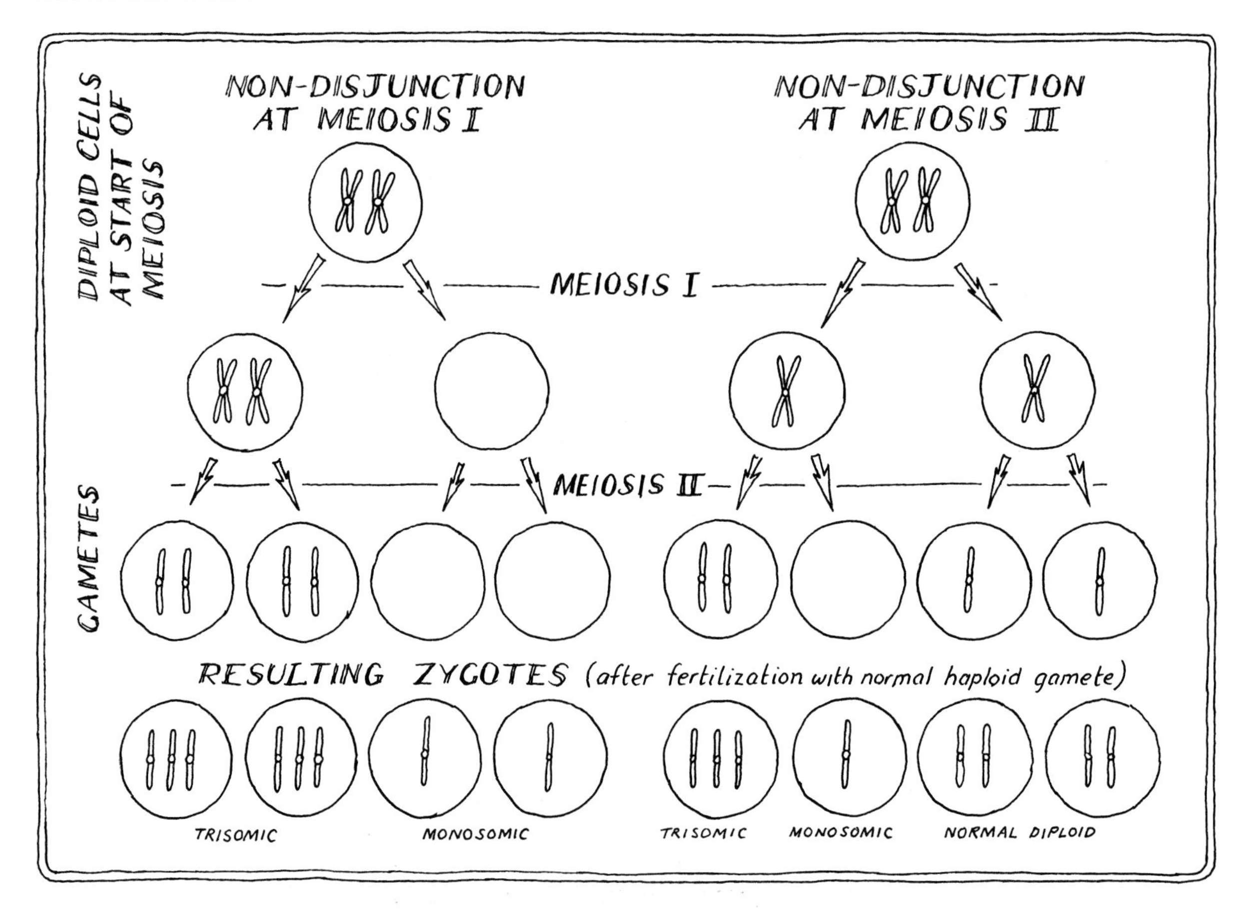

8.3 Monosomy

The phenotypic consequences to the individual who develops from an aneuploid zygote varies depending upon the chromosome involved, but may be severe. Throughout the animal and plant kingdoms the consequences of either having an extra sex chromosome, or lacking one, are less severe compared with changing the number of autosomes. Indeed monosomy for the X chromosome is relatively common. About 1 in 3000 female humans show **Turner syndrome (XO)**. Full sexual development fails to occur in these individuals, who have a single X chromosome; ovaries are rudimentary, although external genitalia and internal ducts are present. All other systems develop normally. By contrast monosomy for one of the autosomes is generally lethal in animal species. It is better tolerated in the plant kingdom, although monosomic plants usually show reduced viability compared to their normal diploid counterparts.

So why should monosomic individuals fail to survive, and why is there a difference in toleration of monosomy between animals and plants? The normal situation is for individuals, being diploid, to possess two alleles for each gene. There must, therefore, be many heterozygous situations in which a functional dominant allele is masking a potentially lethal recessive one. Monosomic cells have an unpaired chromosome. If a lethal allele is present at just one locus on this chromosome, then early death of the developing organism occurs. This explanation would account for the observation that 95% of conceived XO (Turner syndrome) individuals fail to develop. We can assume that the remaining 5% have no detrimental recessive alleles on their unpaired X chromosome and so, except for sex-related traits, development can be normal.

8.4 Trisomy

In contrast to monosomy, the phenotypic consequences of an extra chromosome are not as severe. In plants trisomic individuals are often fully viable, although their phenotype may be altered when compared with normal diploid individuals. A classic example of trisomy involves the much studied Jimson weed, *Datura stramonium*. The diploid number of this species is 24. Twelve different varieties are available and each is trisomic for a different chromosome. Interestingly, each trisomy sufficiently alters the development of the seed capsule to produce 12 different forms (Fig. 8.3).

The viability of trisomic animals varies and seems to relate to the size of the chromosome represented three times, and therefore the number of genes, involved. Thus, it is rare to find individuals trisomic for the larger chromosomes. An exception again to this rule is the X chromosome. Males with an extra X chromosome are found in many mammalian species. About 1 in 500 human males are **XXY**, i.e. show **Klinefelter syndrome**. Their main difference, compared to XY males, relates to ambiguous sexual characters. The presence of the Y chromosome ensures XXY individuals are phenotypically male. Secondary female sexual development is not, however, totally suppressed; so, for example, slight enlargement of the breasts is common.

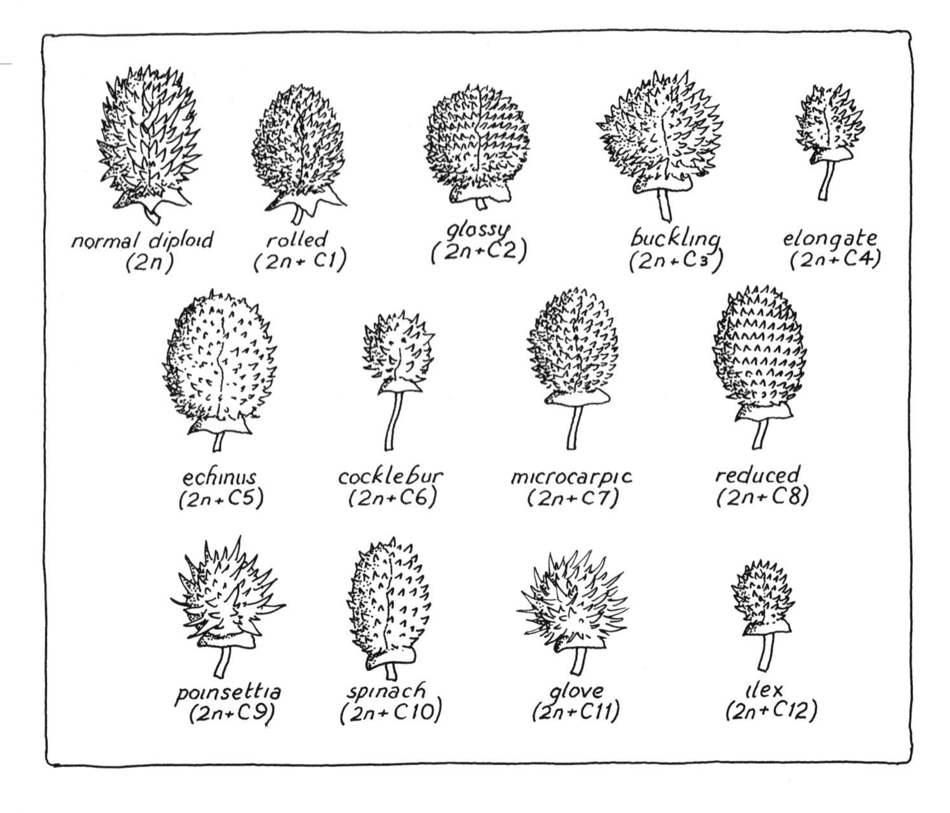

Fig. 8.3
Seed capsules of trisomic varieties of Jimson weed.

These problems are accentuated in individuals who occasionally possess more X chromosomes, i.e. XXXY or even XXXXY!

The only human autosomal trisomy in which a significant number of individuals survive longer than a year is trisomy of chromosome 21, also called **Down syndrome** (Fig. 8.4). Given its relatively high frequency (approximately 1 in 700 live births), it is a condition with which most of us are familiar (Box 8.1). Only two other human trisomies allow individuals to survive to term – trisomy 13 (**Patau syndrome**) and trisomy 18 (**Edwards syndrome**). Both syndromes result in severe malformations and early death.

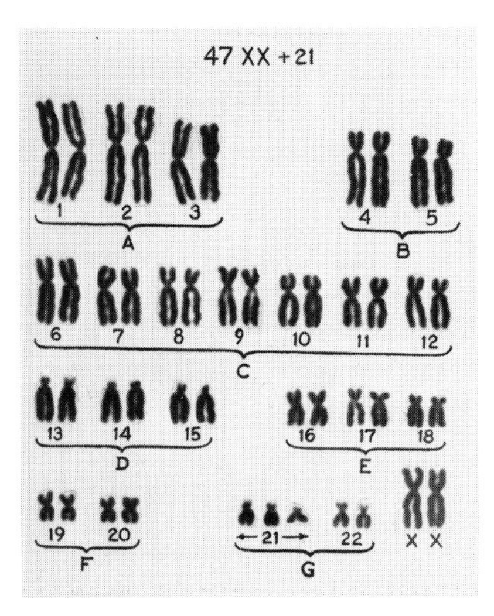

Fig. 8.4
Karyotype of an individual with Down syndrome.

Trisomies of all other human chromosomes, except chromosome 1, are known, but the affected foetuses are spontaneously aborted. Trisomy of chromosome 1 probably produces such profound changes to the human phenotype that death occurs at a very early embryological stage, and so it is not detected. Chromosome 1 is one of the largest human chromosomes with, therefore, a large number of genes to be affected by the imbalance in chromosome complement.

It would appear, therefore, that normal eukaryotic development requires a precise diploid complement of chromosomes. Losing or gaining chromosomes so severely disrupts the delicate balance that normal development is often impossible. Yet, as was illustrated by the Jimson weed, this is not necessarily the case in plants. Trisomies, in particular, can result in useful variants. Indeed, it can be even more positive for plants to have extra sets of chromosomes, i.e. to be **polyploid**.

Box 8.1 Features of Down syndrome

We may recognize someone as a Down individual as they possess a distinctive phenotype. They are usually of small stature; their eyes are narrow because of a prominent fold of skin in the corner of their eyes; their tongues are large and they have broad hands with characteristic palm and fingerprints. Indeed the effect of having an extra chromosome 21 is wide-ranging. Physical, psychomotor and mental development are all retarded. Down people are also prone to respiratory disease and heart malformations and show a higher incidence of leukaemia. Death is frequently due to Alzheimer's disease. This correlates with the recent discovery that a gene promoting early onset Alzheimer's disease is located on chromosome 21.

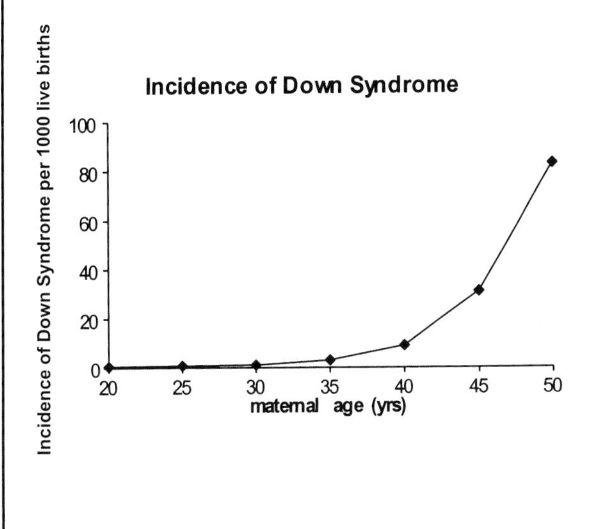

As with all chromosomal abnormalities, the incidence of Down syndrome increases with maternal age. The reason is unknown, but many scientists would relate the dramatic increase that occurs with advancing maternal age to the greater age of the ova. All a woman's potential eggs are formed by birth, and arrested at meiosis I. An older unused egg has had a longer period in which non-disjunction and other defects can occur.

It is routine for older women to be offered prenatal screening for a Down foetus. Foetal cells are obtained by **amniocentesis** or **chorionic villi sampling** and a karyotype prepared. If three copies of chromosome 21 are present the prospective parents can decide whether or not to continue with the pregnancy. All pregnant women now have the opportunity of a blood test at week 16 to assess the levels of three substances. If human chorionic gonadotrophin levels are raised and levels of oestriol and alpha fetoprotein lowered this indicates an increased risk (1 in 250 or greater) of the foetus being trisomic for chromosome 21 and further tests are offered.

Although we can detect the presence of an extra chromosome 21 we cannot predict how severely affected any Down child will be. Many Down people lead happy productive lives, while others are severely mentally and physically handicapped.

develops to term. However, all systems are so profoundly affected that the baby cannot live more than a few hours or days.

3. An autotriploid is produced by mating a tetraploid with a diploid individual. The latter produces haploid gametes while the tetraploid forms diploid gametes. Three sets of chromosomes will therefore exist in the zygote.

8.8 Producing an allopolyploid

Allopolyploidy results from the hybridization of two closely related species and subsequent chromosome duplication. Consider Fig. 8.6. Species A with three pairs of chromosome crosses with species B with a different set of three chromosome pairs. The resulting hybrid is diploid and sterile. Viable gametes cannot be produced as the chromosomes from the two different species cannot pair during meiosis. The hybrid's propagation relies solely upon asexual reproduction. If, however, the hybrid undergoes a natural or induced chromosomal doubling, a fertile **allotetraploid** or **amphidiploid** (two complete diploid genomes) is produced. There are now two homologous sets of chromosomes. The chromosomes of each separate set pair during meiosis. Normal segregation occurs at anaphase and balanced viable gametes are produced.

Fig. 8.6
The origin of an allotetraploid.

	Species A	Species B
Chromosome pairs	a_1a_1 a_2a_2 a_3a_3	b_1b_1 b_2b_2 b_3b_3
Gametes	a_1 a_2 a_3	b_1 b_2 b_3

Sterile diploid a_1 a_2 a_3 b_1 b_2 b_3

chromosome doubling

Fertile tetraploid a_1a_1 a_2a_2 a_3a_3 b_1b_1 b_2b_2 b_3b_3

There are some well documented and commercially important examples of alloploidy. Modern cultivated wheat, *Triticum aestivum*, is an allohexaploid, and is believed to be the result of two natural hybridizations and chromosomal doubling events (Fig. 8.7). Cultivated cotton, *Gossypium* (various species) is a natural hybrid between an Old World and a wild American species in which colchicine was used to induce a chromosome doubling. In both cases the polyploid form is considerably more prolific than the original diploid species. The commercially useful *Triticale* is a designer hybrid between emmer wheat (*Triticum turgidum*) and rye (*Secale cereale*). The aim was to combine wheat's high yielding character and high protein content of its grain with the flexible growth requirements and high lysine content of rye grains.

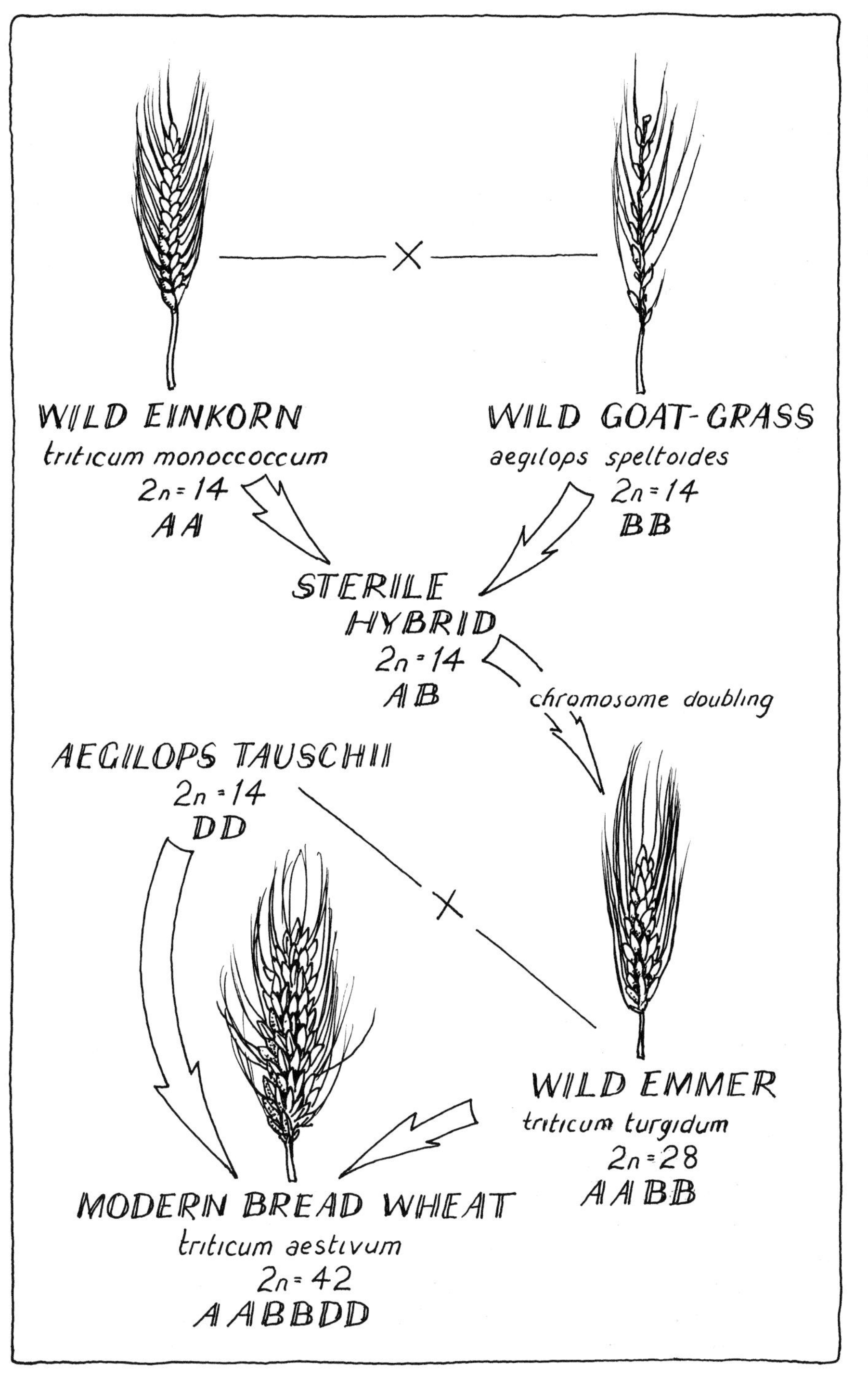

Fig. 8.7
The wheat story: probable evolutionary history of wheat over the last 10,000 years. The letters A, B and D represent different haploid sets of chromosomes. In many modern varieties the long hairs have been eliminated by selective breeding.

8.9 Changes in chromosome structure

The second group of chromosome aberrations that can result in changed phenotypes are structural changes to individual chromosomes. Portions may be deleted or added, or a rearrangement of genetic material may occur, as illustrated in Fig. 8.8.

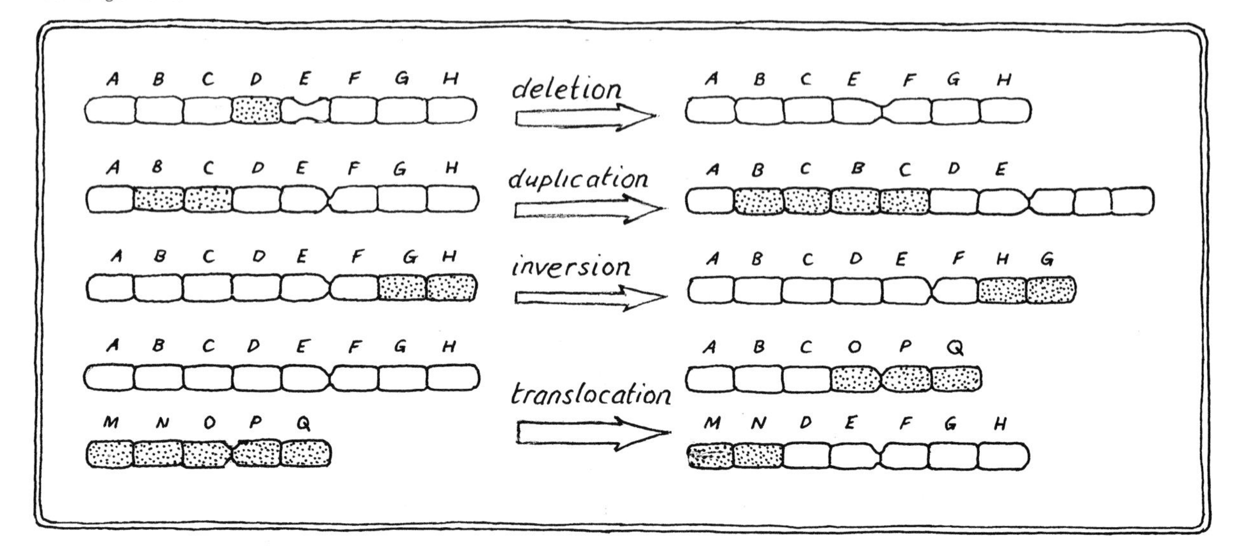

Fig. 8.8
Summary of chromosomal rearrangements.

These changes occur because of one or more breaks along the axis of a chromosome. Chromosomal breakage can occur spontaneously and is, of course, intrinsic to the process of crossing over during prophase I of meiosis. Breakage rates are increased in cells exposed to certain chemicals or radiation. A broken piece of chromosome readily joins with another broken end. If a breakage and rejoining event does not re-establish the original relationship and, if the alteration occurs during gamete formation, the gametes will contain the structural rearrangement. The changed chromosome will be heritable and may have major effects on an individual's phenotype.

The effect of a chromosomal rearrangement may be immediate or delayed for a generation. The timing of expression is related to the type of structural change. If, for example, genes are missing because of a deletion or a crucial gene's function is abolished because a break point occurred within the gene, then the effect is immediate. If, however, no loss or gain of genetic information occurred then the phenotypic consequences of a rearrangement may not be felt by subsequent generations. These phenotypic changes are a consequence of problems during meiosis. Many rearrangements cause problems for homologue pairing during prophase I. These result in gametes either deficient in, or containing extra, chromosomal material. Individuals resulting from these gametes have an increased probability of showing multiple phenotypic changes.

Chromosomal rearrangements in somatic cells are increasingly being linked to the development of human cancers. It is difficult to know whether the various chromosomal aberrations observed in tumour cells are the cause of the tumour or whether they result from the changed growth activities of the tumour cells. There is, however, growing support for the first hypothesis. In some cancers the associated chromosomal change is always the same. For example, in chronic myelogenous leukaemia and in Burkett's lymphoma the same translocation is consistently detected (Table 8.2). In each case the translocation has been shown to activate an **oncogene**, i.e. a gene whose product initiates the transformation from a normal differentiated cell to one capable of the uncontrolled growth characteristic of tumour cells.

Table 8.2
Chromosomal changes and human cancers

Cancer	Chromosomal change
Chronic myelogenous leukaemia	Translocation of long arm of chromosome 22 to chromosome 9
Burkett's lymphoma	Translocation between chromosomes 8 and 14
Ovarian papillary carcinoma	Translocation between chromosomes 6 and 14
Mixed parotid carcinoma	Translocation between chromosomes 3 and 8
Neuroblastoma	Deletion of end of short arm of chromosome 1
Melanoma	Deletion of end of short arm of chromosome 1
Small cell lung carcinoma	Deletion of section of short arm of chromosome 3
Wilm's tumour	Deletion of end of short arm of chromosome 11
Testicular tumour	Deletion of big section of chromosome 12

8.10 Changes in the arrangement of genes

An **inversion** occurs when a segment of a chromosome is rotated by 180°. This is believed to occur when a chromosome forms a tight loop and breaks in two places. The segment between the break untwists slightly and the ends rejoin wrongly, as shown in Fig. 8.9.

The inverted segment may be of variable length. It may or may not include the centromere. As long as a breakage did not occur within a gene there is usually no effect on phenotype. An individual with an inversion may, though, produce aberrant gametes.

Another type of rearrangement happens when breakage occurs simultaneously in two different homologous chromosomes and they exchange segments. This is a **translocation**: a segment is moved to a new position within the genome (Fig. 8.10).

Again, unless breakage abolishes a critical gene function, the presence of a translocation does not directly alter the viability of an individual. However the presence of a translocation causes pairing problems during meiosis. The result is a proportion of gametes that are unbalanced, i.e. they contain chromosomes with duplicated and deleted segments. When used in fertilization the resulting individual may be partially trisomic or monosomic for a chromosome. Being partially trisomic for chromosome 21, because of a translocation, is another way that a Down individual arises (Box 8.2).

Fig. 8.9

Production of a chromosomal inversion.

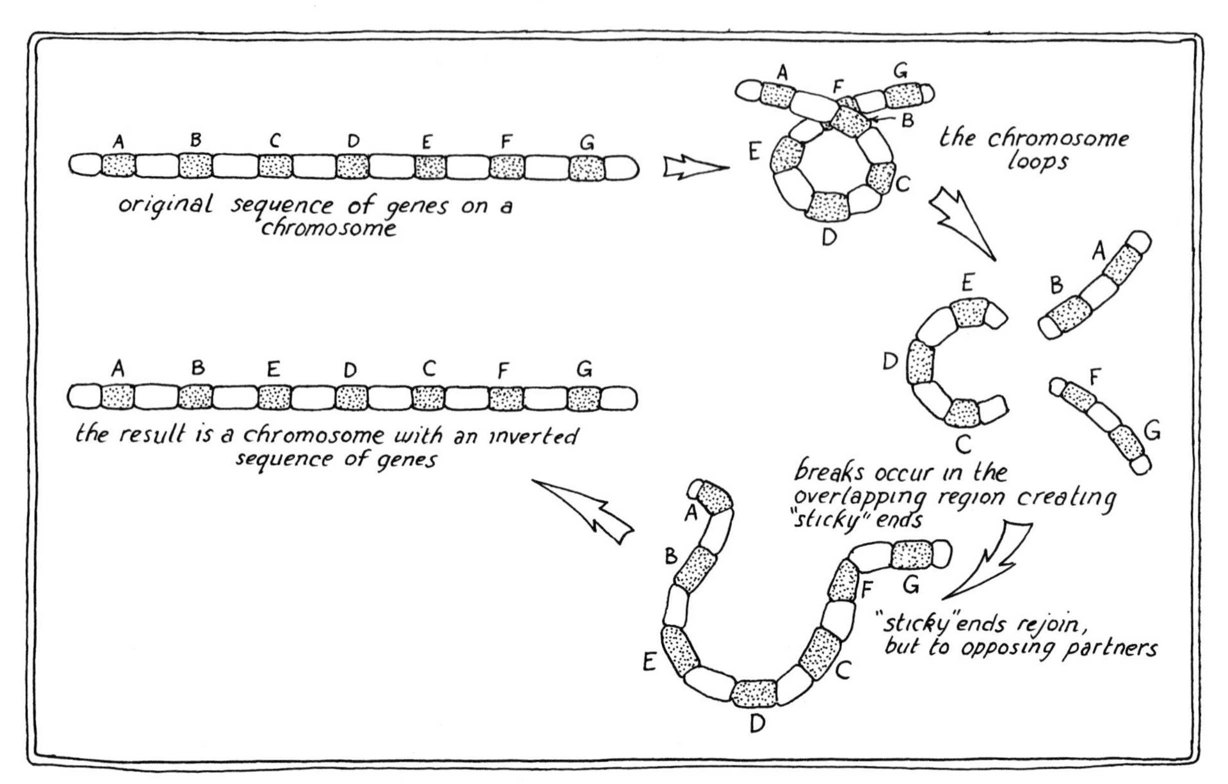

Fig. 8.10

Origins of a reciprocal translocation.

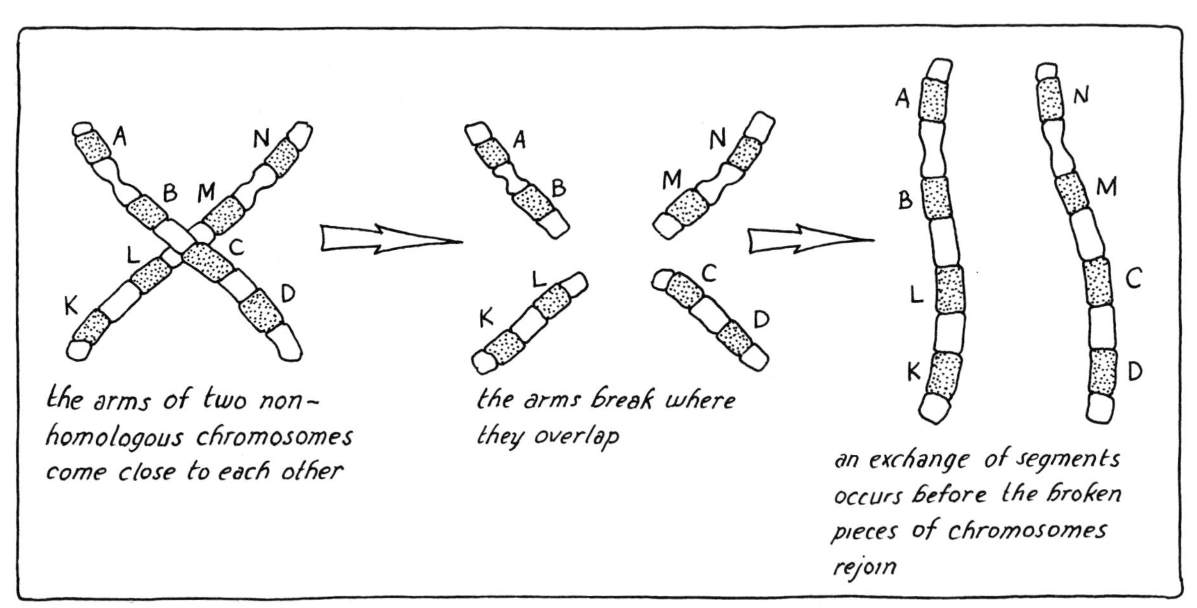

Box 8.2 Production of familial Down syndrome

Most translocations are **reciprocal**: with portions exchanged. Another kind involves break points at the extreme ends of the short arms of two non-homologous chromosomes. The small segments are lost and the longer ones fuse to produce one new large chromosome. This is a **Robertsonian translocation**. It can occur between chromosomes 14 and 21.

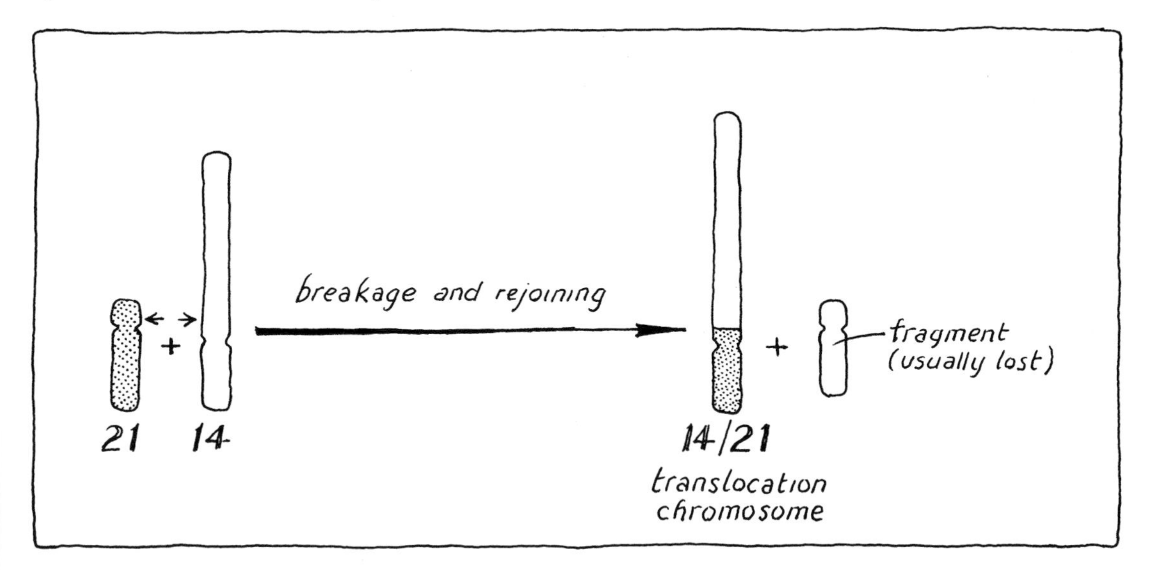

This explains the 5% of cases where Down syndrome is inherited or familial. In these cases one of the parents has a 14/21 translocation. He or she is phenotypically normal, even though there are only 45 chromosomes, as all the relevant genetic information is present. However, when the individual produces gametes, one-quarter of the gametes will have two copies of chromosome 21. When this is fertilized by a standard haploid gamete the resulting zygote, although having 46 chromosomes, is trisomic for chromosome 21, and so will express Down syndrome.

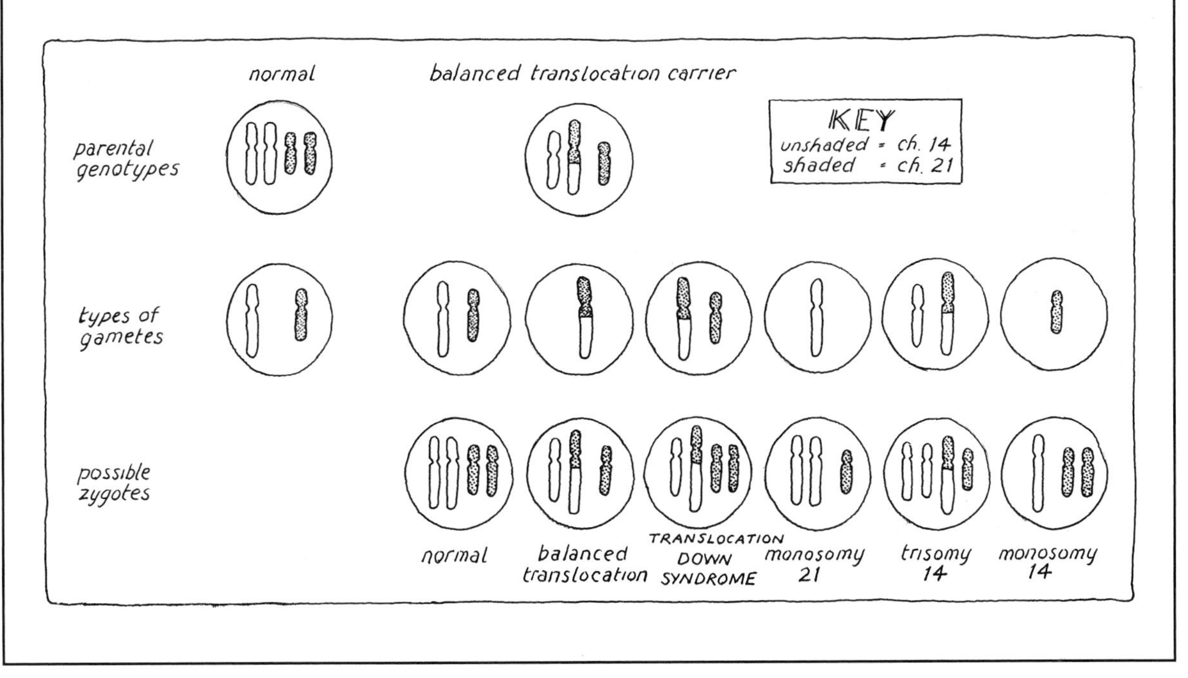

8.11 Changes in the number of genes on a chromosome

Single genes or large pieces of a chromosome may be gained (a **duplication**) or lost (a **deletion**). The commonest cause is unequal crossing over between homologues during meiosis, so that one chromatid gains, and the other loses, a region. This can occur when homologous chromosomes mispair – see Fig. 8.11.

Fig. 8.11

The origin of chromosomes with duplicated or deleted regions.

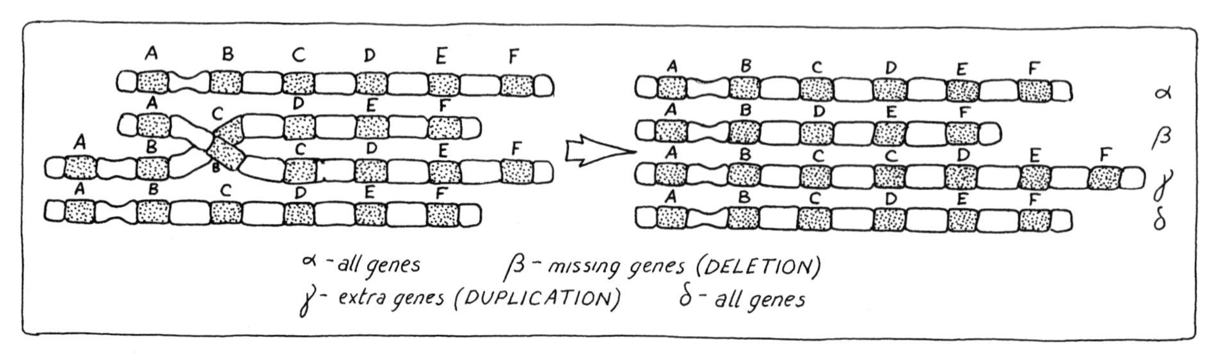

If a gamete carrying a deletion or duplication is viable and fuses with a normal gamete, the individual produced may show a new phenotype; for example, the *Cri du Chat* (French, 'cry of a cat') syndrome in humans or the *Bar* eye phenotype of the fruit fly *Drosophila melanogaster. Cri du Chat* syndrome is caused by a deletion of much of the short arm of chromosome 5, and is characterized by mental retardation, microcephaly (a small head) and a distinctive cry. *Bar* eyed flies have a narrow slit-like eye instead of the normal oval-shaped one (Fig. 8.12), and duplication of a piece of the X chromosome. The duplication is inherited in a dominant fashion.

Fig. 8.12

Gene duplication and the production of the *Bar* eye phenotype in *Drosophila melanogaster.*

genotype	number of eye facets	eye phenotype (shape and size)	region 16A of X chromosome
bb	800		
Bb	350		
BB	70		

It is obviously disadvantageous for the fruit fly to have *Bar* eyes as, with fewer eye facets, its vision is impaired. Gene duplication can, however, play a positive role in cells. Some gene products are needed in large amounts in cells. These genes are present in multiple duplicated copies, e.g. the gene coding for rRNA which supports protein synthesis. Gene duplication is believed to have played an important role in evolution. Pairs of genes have been found that have a substantial amount of their DNA sequence in common, for example the protein digestion enzymes trypsin and chymotrypsin; also the oxygen carriers myoglobin and haemoglobin. Such gene pairs are believed to have arisen from a common ancestor through duplication. The related genes diverged sufficiently during evolution to produce unique products.

Summary

- If the chromosome complement of a diploid organism is changed it can have profound phenotypic consequences.
- A polyploid individual possesses one or more extra sets of chromosomes while the cells of an aneuploid individual lack or contain extra single chromosomes.
- Non-disjunction of chromosomes or chromatids during meiosis accounts for the production of aneuploids.
- Monosomic individuals lack one chromosome. In animals this is usually lethal, but it is better tolerated in plants.
- Trisomic individuals possess an extra chromosome. This results in variable viability in animals. In plants trisomy can produce useful variants.
- Polyploidy is a widespread phenomenon in plants. Autopolyploids possess multiple sets of the same chromosome while allopolyploids have chromosome sets from two or more different species.
- Portions of a chromosome may be deleted or duplicated or inverted by 180° or exchanged with segments from another chromosome. These structural changes to chromosomes can have a variety of different consequences depending upon the nature of the change and the stage of development when they occur.

Problems

1. In humans there are 23 chromosomes per haploid set. Aborted foetuses were found to have the following conditions. How many chromosomes would you expect in the somatic cells of each of the following individuals?

   (a) monosomy 3 (b) triploid

   (c) trisomy 16 (d) tetrasomy 14

2. Below is the normal gene sequence on a chromosome:

   HIJKL.MNOPQ

   What chromosome change resulted in the following chromosomes?

   (a) HIJKNM.LOPQ

   (b) HIJKL.MNONOPQ

   (c) HIJKLLK.MNOPQ

   (d) HIJKL.MNCDEFG

3. A diploid species with 32 chromosomes is crossed with another with 26 chromosomes.

   (a) What would be the number of chromosomes in the hybrid, and in the allopolyploid derived from this hybrid?

   (b) Would you expect the hybrid and polyploid offspring to be fertile?

4. Only rarely does an inversion or a reciprocal translocation have a detrimental effect upon the phenotype of an individual. Explain why.

5. What types of gametes, and in what proportions, would an individual trisomic for a chromosome produce?

6. Genes PQRSTU are known to be closely linked, but there are doubts as to their correct order. A set of individuals with deletions in this area were obtained. The deletions uncover recessive mutations:

   deletion 1 uncovers p, q and s

   deletion 2 uncovers p, s, r and u

   deletion 3 uncovers t and u

   deletion 4 uncovers q and s

   What is the order of these genes?

7. Two phenotypically normal parents have a child with severe learning difficulties. A karyotype revealed that the child had two normal copies of chromosome 12 but only one normal copy of chromosome 9. The second copy of chromosome 9 had a large section of its short arm replaced by a piece of chromosome 12. Suggest, with reference to chromosomes 9 and 12, the chromosome compositions of the child's parents.

8. When mitosis was examined in root tip cells from an autohexaploid plant species, each cell was found to contain 72 chromosomes. How many chromosomes did each gamete of the original diploid plant contain?

9. An allotetraploid plant has 44 chromosomes in each somatic cell. How many linkage groups are there?

10. A woman with Turner's syndrome also shows the X-linked condition, red-green colour blindness. She had a colour blind father and a mother with normal vision. In which of her parents did non-disjunction of the sex chromosomes occur?

11. How many chromosomes would be found in the somatic cells of an allohexaploid plant derived from three species in which the haploid chromosome number is 9, 11 and 12, respectively?

9 The principles of quantitative genetics

A farmer moved from a fertile plateau at an altitude of 2000 metres in the Andes to a new farm, much lower on the Patagonian plains, taking with him a stock of his high-yielding prize potatoes. He confidently looked forward to a good crop at the end of his first year. Instead, he was dismayed by the yield of few and small potatoes. Attributing the low yield to the inevitable, but occasional, bad year, he planted out his seed potatoes again the following year, only to obtain an even poorer harvest. The farmer was bemused. He was aware of the principles of Mendelian genetics and so had expected that the high-yielding gene, or genes, that his potatoes possessed would continue to be passed to subsequent generations, and thus ensure him

Fig. 9.2

Histogram showing the variation in tree height in Mikumi National Park.

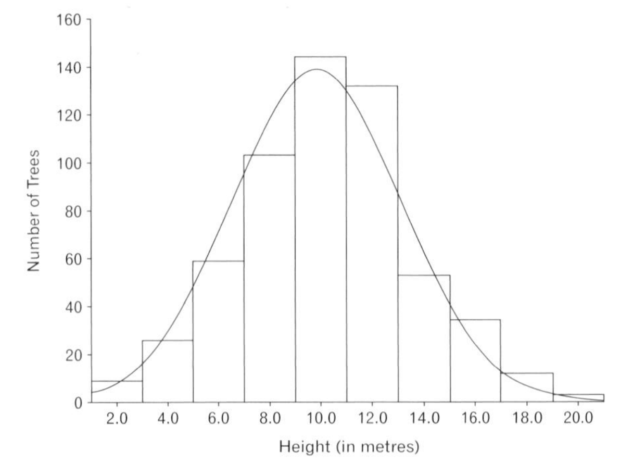

- 576 trees were measured. This produced a height range from 1.8 to 24.6 metres.
- The height range was divided into ten equally spaced classes.
- The number of trees that fell into each class was counted.
- The data was plotted with the numbers in each class on the vertical axis and the class sizes on the horizontal axis (Fig. 9.2).

It can be seen that the histogram representing the range of tree heights is bell-shaped; or stated more formally, shows a **normal distribution**. The height of the majority of trees fell within the middle of the range with numbers falling off towards both ends. The frequency diagram for most multifactorial traits approximates to a normal distribution. This important observation means that we can use the properties of a normal distribution to investigate the variation shown by multifactorial traits; and thus come up with an estimation of the relative contribution of genetic and environmental factors to phenotype.

From a sample of measurements, such as tree heights, we can calculate:

- the **mean**, or arithmetic average;
- the **variance**, which is an indicator of the spread of the measurements around the mean, i.e. whether the sample shows a small or large range of variation;
- the **standard deviation**, which is another indicator of the range of variation shown by the sample.

9.3 Calculating the mean

A normal distribution shows that a set of measurements tends to cluster around a central value. The mean ($\bar{x}$) is a measure of this central value. It is calculated by summing together each individual measurement and dividing the resulting value by the number of measurements:

$$\bar{x} = \frac{\sum x}{n}$$

$\sum$ = sum of (i.e. all values added together); x = individual values; n = sample size

Box 9.1 Calculating mean elephant tusk weight

The table on the right shows the calculation of a mean tusk weight found on dead elephants in Mikumi National Park, Tanzania (Fig. 9.3). Each elephant has (generally) a pair of tusks. Thus the average weight of the two tusks of a pair was first calculated and then the average weight of a tusk for all 12 elephants was determined. This was achieved by summing the weights of all the tusks and dividing the summed value by 12 (the sample size) to give an overall mean tusk weight.

Elephant	**Weights of individual tusks of a pair (kg)**		**Average weight of a tusk in a pair (kg)**
1	21.00	21.40	21.20
2	1.20	1.20	1.20
3	1.00	1.00	1.00
4	0.70	0.60	0.65
5	4.00	3.75	3.88
6	7.00	7.50	7.25
7	9.00	9.00	9.00
8	2.10	1.90	2.00
9	1.03	1.02	1.03
10	1.02	1.02	1.02
11	7.20	7.00	7.10
12	1.90	2.10	2.00
			$\sum x$ = **58.33**

$$\bar{x} = \frac{\sum x}{n} = \frac{58.33}{12} = 4.86$$

Thus the average weight of a tusk found on a dead elephant was 4.86 kg.

The mean measurement of a sample is, by itself, of limited value as it does not tell us anything about the spread of our results or, in a genetic context, how variable is the expression of a trait. For example the tusk weights presented in Box 9.1 are very variable. Furthermore it is possible for two sets of measurements to share the same mean value but show a great difference in their ranges. Hence it is important to also have a value for the **variance**.

Fig. 9.3

A family group of elephants in Mikumi National Park, Tanzania: members of the largest remaining population of open country elephants in Africa, stretching through the adjacent Selous Game Reserve and beyond. The consequences of lifting the CITES (Convention on International Trade in Endangered Species) trade ban on ivory is being closely monitored.

9.6 Genes or environment?

The variance gives an indication of the range of phenotypic expression for a given trait in the particular population under study. Within quantitative genetics we refer to this as **phenotypic variance** (V_p). The next task is to assess how much of the variation represented by the phenotypic variance might be caused by genetic factors and how much caused by environmental ones. We therefore attempt to break down, or **partition**, the phenotypic variance into two values, a **genetic variance** (V_g) and an **environmental variance**(V_e). Thus:

phenotypic variance = genetic variance + environmental variance

$$V_p = V_g + V_e$$

The approach to estimating separate genetic and environmental variances is essentially the same in all plants and non-human animals. If individuals of uniform genotype are nurtured under a range of different environmental conditions, any phenotypic variation should be attributable to environmental factors. The variance calculated from these measurements should, therefore, be an estimate of **environmental variance**. Similarly, if individuals of varying genotype develop under the same set of environmental conditions, any phenotypic variation should be the result of genetic factors and give us a value for the **genetic variance**. Necessarily, any values can only be an estimate, because how can we be certain that a group of individuals are subjected to exactly the same environmental conditions or, given the polygenic nature of most multifactorial traits, how can we be sure that individuals are homozygous for all the relevant genes? There could also be confounding variables such as different genotypes behaving differently in different environments! However, these approaches to controlling genotype of the environment can be used to produce values for the two different variances and so provide useful estimates of the genetic input to various quantitative traits in a wide variety of plants and animals, but not in humans. Because these approaches involve setting up specific genetic crosses or carefully controlling the environment, other methods obviously have to be used when assessing the relative contribution of inheritance and the environment in humans. These are discussed in Section 9.11. Interest in this question is, however, high with regard to humans, because it forms part of the hotly debated 'nature–nurture' contribution to many of our human traits.

9.7 Partitioning phenotypic variance

A pseudo-Mendelian approach is adopted as the basis for calculating genetic and environmental variance. Individuals from two pure-breeding populations are crossed. A value for the variance in the parental, F_1 and F_2 generations is calculated. From this data it is possible to partition overall, phenotypic, variance into genetic and environmental variances. The principles are best demonstrated by using some data.

Consider some data collected in an early study of quantitative traits. The American plant geneticist Edward East (1879–1938) examined the inheritance of corolla tube length in several strains of the tobacco plant (*Nicotiana longiflora*). Figure 9.5 shows the variation in length of the corolla tube shown by two pure-breeding parental varieties and by the F_1 and F_2 generations produced by crossing short- and long-flowered plants. A mean and a variance were calculated for each group.

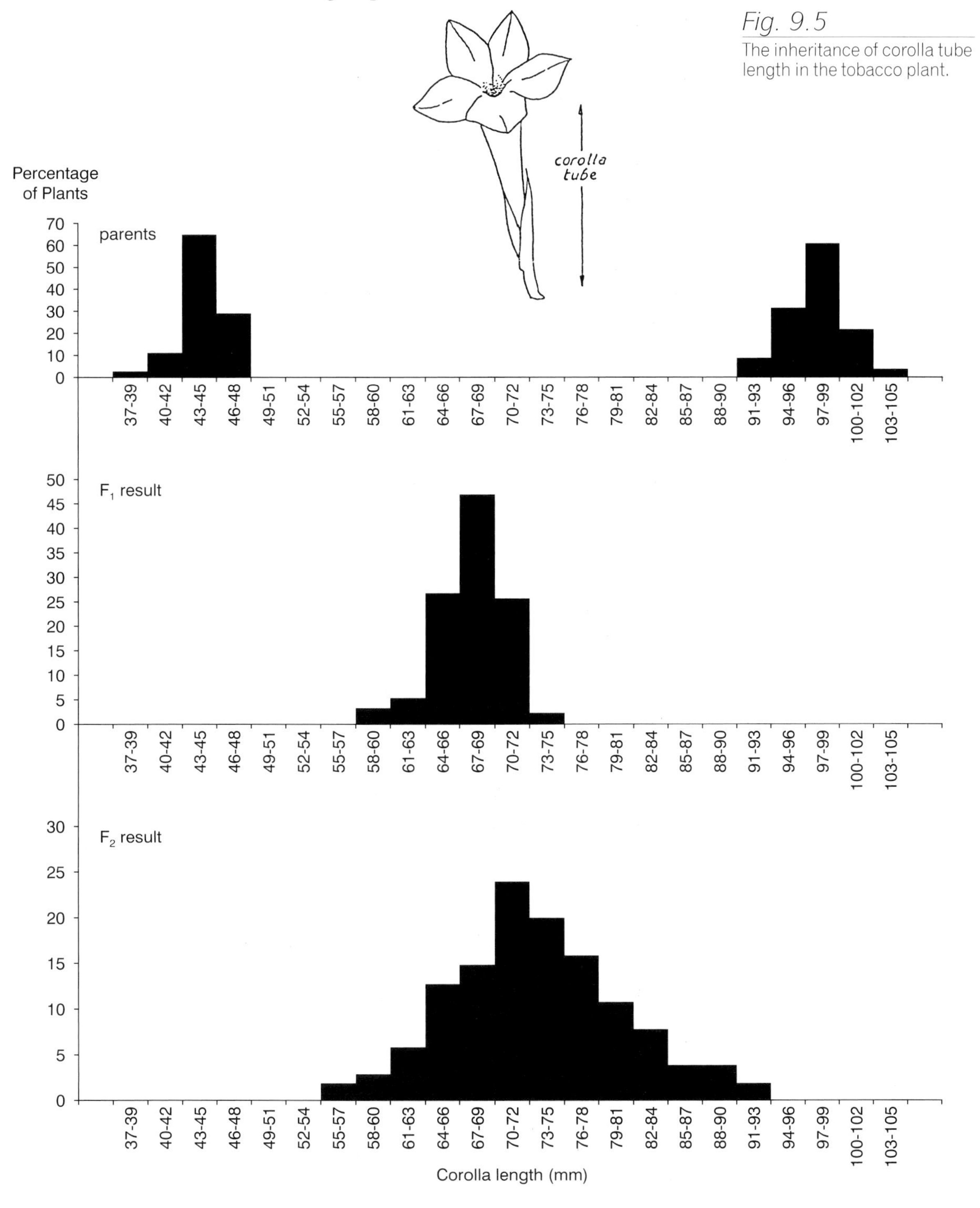

Fig. 9.5
The inheritance of corolla tube length in the tobacco plant.

With reference to the data presented in Fig. 9.5, we can argue that:

1. The variation in corolla length for each parental group of plants results from variable environmental factors because the two varieties were pure-breeding and so should be homozygous for most, if not all, relevant loci. Genetic variance should be, theoretically, zero.
2. The variation among the F_1 offspring should also be solely the result of variable environmental factors because all F_1 individuals should be genetically identical; although this time heterozygous. The expected uniform heterozygosity of the F_1 generation is explained in Fig. 9.6.

Fig. 9.6

Explaining the uniform heterozygosity of the F_1 generation in Fig. 9.5.

Suppose corolla length is the result of the action of four genes – **A, B, C** and **D**

Genotypes of pure-breeding parents	**AABBCCDD**	**aabbccdd**
Genotype of gametes	**ABCD**	**abcd**
Genotype of F_1	**AaBbCcDd**	

3. We therefore have three estimates of environmental variance: one for each parental variety and another for the F_1 generation. For our corolla tube example this gives us an average value of 8.76 for environmental variance.
4. The greater variation in corolla length shown by the F_2 generation results from individuals both experiencing different environmental conditions and possessing different genotypes. When the F_1 generation is crossed among each other many different genotypes will segregate out. The variance of 40.96 therefore represents both environmental variance (V_e) and genetic variance (V_g).
5. Genetic variance, V_g, can therefore be estimated at 32.2. This is calculated using the equation $V_p = V_g + V_e$. We had values for V_p and V_e and so can rearrange the equation to produce a value for V_g:

   $V_p = V_g + V_e$

   $40.76 = V_g + 8.76$

   Rearranging the equation:

   $V_g = 40.76 - 8.76 = 32.2$

   The genetic variance, V_g, is therefore much greater than the environmental variance, V_e. Thus, for this F_2 population, we can argue that the observed variation in corolla length is largely the result of genetic differences between individuals, or, as the next section explains, approximately 80% of the variability in corolla length among the particular population of tobacco plants being studied is attributable to genetic causes.

9.8 Heritability index

The proportion of the total phenotypic variation attributable to genetic differences between individuals is often formally recognized as the **heritability index**. We use values gained for genotypic and phenotypic variance to calculate this index. It is equal to the genotypic variance divided by the phenotypic variance:

$$\text{Heritability index } (H^2) = \frac{V_g}{V_p}$$

For the corolla tube example used in Section 9.7 the heritability index would be:

$$\frac{32.2}{40.76} = 0.79$$

Heritability of a trait ranges from 0 to 1. A heritability index of 0 indicates that none of the variation in phenotype among individuals in a population is attributable to genetic causes; whilst a value of 1 suggests that all the observed phenotypic variation has a genetic basis. Thus, a value of 0.79 for the heritability of our tobacco corolla length example indicates a strong genetic input to the expression of this character in the population under study. Table 9.1 shows heritability values for a selection of different traits in a range of species.

Table 9.1

Heritability estimates for a selection of agriculturally important traits (values are generally averaged for a number of populations in differing environmental conditions)

Trait	Heritability
Egg weight in poultry	0.55
Egg production (to 72 weeks) in poultry	0.10
Age at first laying	0.52
Litter size in pigs	0.05
Fleece weight in sheep	0.40
Milk production in cattle	0.33
Plant height in maize	0.70
Ear length in maize	0.17
Yield in maize	0.25
Height in spring wheat	0.72
Tiller number in winter wheat	0.36

At first sight a value for the heritability index would seem an extremely useful thing to know. Consider a dairy farmer, keen to increase the milk yield from their herd. They need to know how much of the current variability in milk production by their cows is environmentally induced and how much is a result of genetic factors. If the farmer had access to a heritability index value for milk production then they would surely know whether a selective breeding programme is likely to be successful in raising milk yields, because a high value indicates a strong genetic component and vice versa. Great care, however, has to be exercised when using a heritability index.

Any heritability index is sample- and circumstance-specific. The variances used to calculate a heritability index are for the particular population under investigation and the conditions prevailing at the time the measurements are made. A heritability index therefore represents how much of the observed variation in phenotype is attributable to genetic factors under the specific set of prevailing conditions. It is impossible to obtain an absolute value of the heritability index for a given trait. However, over the years, a sufficient number of estimates of the heritability index have been made for certain traits for a general picture to emerge as to the relative contribution of genetic and environmental factors to expression of these traits.

There are further reasons for caution in using the heritability index as a measure of the potential success of a breeding programme. Consideration also has to be given to the different types of genetic factors that influence phenotype. The heritability index we have so far been discussing is often referred to as **broad sense heritability**, and is really too broad to be useful. This is because it ignores the fact that the genetic contribution to the expression of a phenotype encompasses different types of gene action and interaction. Genetic variance itself can be partitioned into different components. To know about one of these, the **additive genetic variance**, is often of more use to the plant or animal breeder.

9.9 Partitioning genetic variance

Genetic variance (V_g) can be split into the following three major subcomponents:

- **additive genetic variance** or V_a
- **dominance variance** or V_d
- **interactive variance** or V_i

So:

$$V_g = V_a + V_d + V_i$$

These three subcomponents reflect the different ways in which gene action and interaction can contribute to the phenotype of an individual. It is important, for example, to know whether alleles at different contributing loci show complete or incomplete dominance (reflected in the **additive** and **dominance** variances), and the extent to which alleles at different loci interact to bring about expression of a particular phenotype (reflected in the **interactive** variance). These three different types of genetic variances are discussed in more detail in the rest of this section.

The **additive genetic variance** relates to alleles at a locus being incompletely dominant. Such alleles influence phenotype, such as height, colour or mass, in an additive or quantitative fashion. Consider the genetic basis of snapdragon flower colour in relation to pigment levels in the petals (discussed in Section 3.2). The three main colours of red, pink and white are determined by two incompletely dominant alleles at one locus. The relationship between genotype and phenotype is summarized in Table 9.2.

Genotype	Pigment units	Petal colour
WW	None	White
WR	One	Pink
RR	Two	Red

Table 9.2

Relationship between genotype and snapdragon petal colour in terms of pigment units

If the same **additive** principle applies to the situation where a number of genes contribute to the expression of a particular trait then a wide range of different phenotypes results from the different genotypes (and this is before environmental factors are taken into consideration!). Consider a theoretical situation where two genes, **A** and **B**, control wheat yield (it is actually controlled by many more genes), and where there are two alleles at each loci that contribute to yield in the following relative manner:

A allele = 6 units
a allele = 3 units
B allele = 2 units
b allele = 1 unit

Table 9.3 shows the phenotypic distribution in the F_2 generation when pure-breeding **AABB** and **aabb** parents are crossed. It can be seen that *just* with two genes, and two alleles at each locus, there are nine different genotypes and so phenotypes. The range of possible phenotypes is greatly expanded if there are more genes contributing to expression of a trait, and especially so if at least some of these genes possess multiple alleles.

Genotype	Ratio in F_2	Phenotype (yield units)
AABB	1	16
AABb	2	15
AaBb	1	14
AaBB	2	13
AaBb	4	12
Aabb	2	11
aaBB	1	10
aaBb	2	9
aabb	1	8

Table 9.3

The effect of additive alleles on phenotype

The **dominant genetic variance** represents the variability in phenotypic expression of a trait that is possible by the action of those contributing loci whose alleles show a definite dominant/recessive relationship. Under these circumstances the heterozygous genotype has the same influence on final phenotypic expression as a homozygous dominant one. While there may be considerable genotypic variation in a population, the corresponding phenotypic variation is less.

Finally, epistatic interactions may occur among alleles at different loci contributing to expression of a given trait (discussed in Section 4.6). The existence of epistasis adds another source of variability when considering the influence of genetic differences between individuals upon phenotypic expression of a trait and is represented by the **interactive variance**.

In summary, therefore, the genetic variance (V_g) is partitioned as follows:

$$V_g = V_a + V_d + V_i$$

And the total, or phenotypic, variance (V_p) is partitioned thus:

$$V_p = V_e + V_a + V_d + V_i$$

9.10 Narrow sense heritability

Many commercially important traits are influenced by genes acting in an additive way. Thus, for the farmer wanting to know whether a breeding programme is likely to increase the amount of fleece his sheep produce or reduce the fat content of his meat, it is crucial to know the additive genetic variance V_a. This enables a **narrow sense heritability index** to be calculated:

$$\text{Narrow sense heritability index } (h^2) = \frac{V_a}{V_p}$$

As with the broad sense heritability, a high value indicates that much of the variation in observed phenotype results from differences in genotype at additive loci among individuals in the sample population. In contrast, a low value for the heritability index suggests that environmental factors have a greater influence on expression of the trait under consideration.

It is relatively easy to obtain values for phenotypic, environmental and total genetic variances. It is, however, more difficult to design experiments to estimate additive genetic variance, and thus be able to calculate a value for narrow sense heritability. It is beyond the scope of this book to discuss these calculations at length, although one experimental strategy for obtaining an estimate of the narrow sense heritability will be outlined. It involves performing a one-generation selective breeding programme. If genetic variability for a trait exists among individuals in a population, then it should be possible to change the spectrum of phenotypic expression of that trait by restricting breeding to those individuals which show the desired phenotypes. The extent to which the phenotype changes in one generation is known as the **selection response** (*R*). This can be used to produce an estimate of the narrow sense heritability, as explained in Box 9.3, and thus, an indication as to whether a selective breeding programme will modify the phenotypic expression of a trait in the desired direction.

9.11 Investigating multifactorial traits in humans

It has become increasingly obvious over the last couple of decades that many common human diseases, such as cancer, heart disease, diabetes, schizophrenia and Alzheimer's disease, are the result of a complex interplay between genetic and environmental factors. Thus, there is much current interest in analysing quantitative traits in humans. For various

Box 9.3 Calculating narrow sense heritability

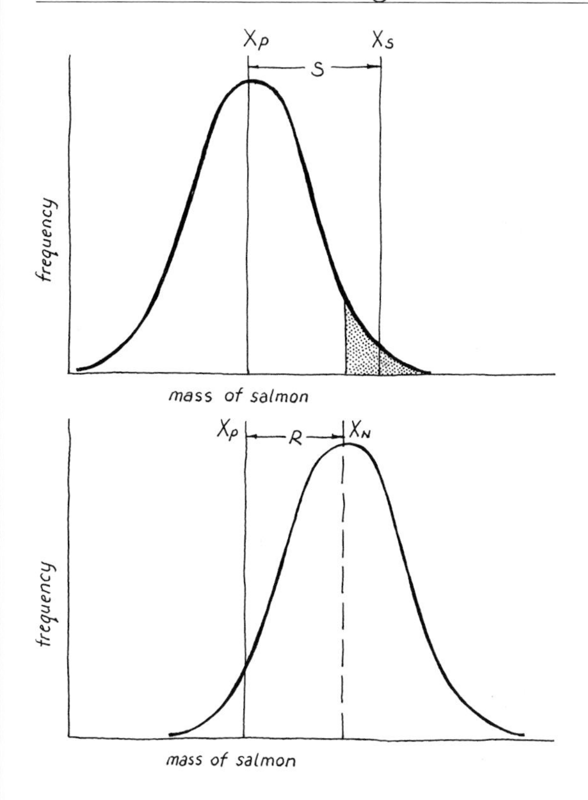

Consider the first graph, which shows the distribution of mass among a sample of 100 salmon on a fish farm. These salmon can be regarded as the parental generation.

Aim: To increase x_P (mean mass of salmon)

Method: Select and interbreed individuals that fall in the shaded section. x_S is the mean mass of the selected fish.

Result: Shown in the second graph: x_N is the mean mass of the offspring.

Selection differential (S) $= x_S - x_P$

Selection response (R) $= x_N - x_P$

Narrow sense heritability $= \frac{R}{S}$

The larger the heritability value, the more likely that a selective breeding programme will be successful.

reasons, however, the approach is different compared to studying such characters in non-human animals and plants. Firstly, as suggested in Section 9.6, traditional heritability studies are not possible in humans. Instead, human twins have proved informative subjects for assessing the relative influence of genes and the environment in the formation of a range of human traits. Secondly, the mode of expression of many human multifactorial traits is different.

Hitherto, the traits we have considered in a range of animals and plants have shown a gradation of expression from, for example, low to high yield in agriculture. A variety of human traits, such as height or hypertension, do show a similar gradation (Fig. 9.7). Many disease conditions are, however, different. They represent an either/or situation: an individual either suffers from or is free of a disease. There is, instead, a gradation of susceptibility or **liability** to developing a disease such as diabetes. Once a **threshold** is reached, the condition is expressed (see Fig. 9.7). The underlying quantitative principle, however, remains the same, in the sense that a number of genetic and environmental factors influence a tendency to express the disease, or some other trait. Any individual who both inherits a particular combination of alleles at various loci and experiences one or more unfavourable environment factors, expresses the phenotype.

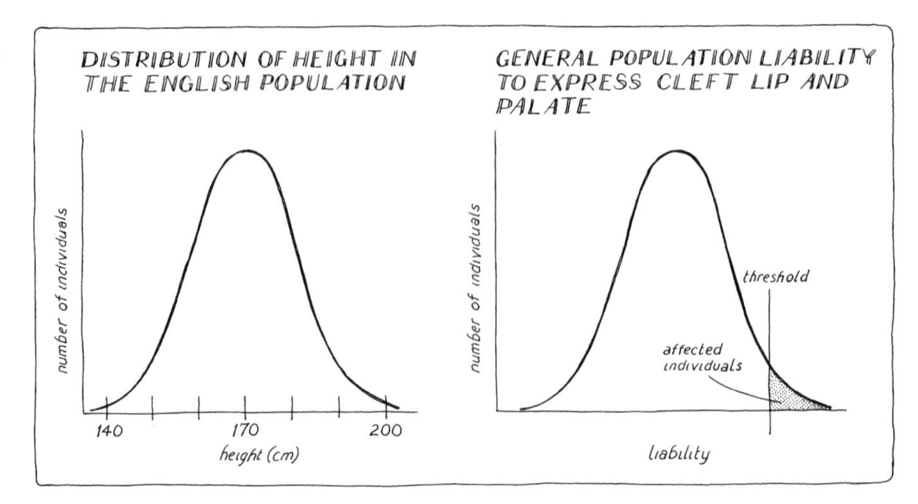

Fig. 9.7
Patterns of expression of a human multifactorial trait: (a) height, (b) liability to develop a cleft lip and palate.

Twin studies have proved invaluable in attempts to measure the relative input of genetic and environmental factors in determining expression of a wide range of human quantitative traits. They involve making comparisons of the frequency of expression of different traits in **monozygous** (or identical) and **dizygous** (or fraternal) twins. Monozygous twins originate when a developing embryo splits to form two separate ones. The resulting individuals are genetically identical, examples of natural clones. Dizygous twins are the result of two separate fertilization events. Thus, as with any two siblings, dizygous twins share just half their genes. Because any pair of twins share broadly the same environment, we are therefore in a position to assess the genetic contribution made to a characteristic. To achieve this, comparisons are made of the frequencies of expression of different traits in monozygous and dizygous twins and **concordance** values are calculated. Twins are said to be:

- **concordant** for a given trait if both or neither of them express it;
- **discordant** if only one shows the trait and the other does not.

For traits totally determined by genotype, concordance rates should be 100% for monozygous twins and 50% for dizygous twins. Values will decrease the greater the environmental input, but always be higher for monozygous twins if genetic factors are involved (Table 9.4). If similarly high concordance values are obtained for both monozygous and dizygous twins this indicates a strong environmental input to the expression of condition, as in the high concordance rates for the development of measles (Table 9.4). The shared environment of both kinds of twins exposes them equally to the infectious agent. As an extension to conclusions drawn from twin studies, any environmental effects can be checked out in the few studies that have been possible on twins reared apart. Do concordance rates remain high when environments are different?

Table 9.4

Concordance rates in monozygous and dizygous twins for selected traits

Trait	Concordance % monozygous twins	Concordance % dizygous twins
Alzheimer's disease	58	26
Cancer	17	11
Cleft palate	35	5
Diabetes mellitus (insulin dependent)	35	6
Diabetes mellitus (non-insulin dependent)	100	10
Epilepsy	37	10
Hypertension	30	10
Ischaemic heart disease	19	8
Leprosy	60	20
Manic depression	70	15
Measles	76	56
Multiple sclerosis	25	6
Pyloric stenosis	15	2
Schizophrenia	45	12
Spina bifida	6	3
Tuberculosis	87	26

There are, however, limits to the information that can be extracted from twin studies. They can never do more than indicate the extent to which a multifactorial condition *may* be caused by genetic factors. If we are to make inroads into understanding and treating some of the major diseases that afflict us, specific genes need to be identified. The application of recent technological advances in molecular biology to the mapping of genes involved in the expression of multifactorial traits, or **quantitative trait loci (QTLs)**, are yielding exciting results. For example, a search for genes increasing susceptibility to juvenile onset, insulin dependent diabetes (IDDM), has identified 20 different predisposing loci, two of which produce a strong effect. One of these is a defect in a gene producing a cell surface insulin receptor, and the other is a fault in a gene producing an immunoprotein. Interestingly, IDDM is an autoimmune disease in which the immune system attacks and destroys the insulin-producing Islets of Langerhans in the pancreas. This new era of identifying QTLs produces the real hope that treatment and eventually cure of some of mankind's major killer diseases may become possible. A genetic test early in life might perhaps identify individuals with high susceptibilities to certain diseases, and thus allow the possibility of preventive treatments. Gene therapy might even provide the ultimate cure.

Summary

- The expression of many traits in living organisms depends upon both genetic and environmental factors.
- The relative contribution of genetic and environmental factors to the expression of multifactorial traits is assessed by the use of statistical techniques.
- Measurements are made of the variation shown by a multifactorial trait: a mean, variance and standard deviation is then calculated.
- The phenotypic variance represents the range of phenotypic expression for a given trait. The genetic and environmental variances indicate how much of the variability in expression between individuals of a sample is the result of genetic and environmental factors, respectively.
- The broad sense heritability index represents the proportion of the total phenotypic variation attributable to genetic factors. A high value suggests that a selective breeding programme would be effective in achieving a desired phenotype within a group of individuals. Sometimes the narrow sense heritability index is more informative for this purpose.
- An assessment of the relative input of genetic and environmental factors in determining traits in non-human animals and plants is gained from the results of experiments controlling matings and environmental conditions. In humans such information arises from twin studies.
- Geneticists are beginning to identify genes (QTLs) involved in multifactorial traits.

Problems

1. This question considers a normal distribution:

(a) What term is given to the value along the x-axis that corresponds to the peak of the distribution?

(b) If two normal distributions have the same mean values, but different variances, which distribution is the broader?

(c) What proportions of the population are expected to lie within one, and within two, standard deviations of the mean?

2. The table below represents average blue tit nestling weight for ten nests in a mature oak wood and for ten nests in a marginal scrubland habitat:

Nest	Average nestling weights in woodland habitat (g)	Average nestling weights in scrubland habitat (g)
1	10.4	9.01
2	10.2	9.86
3	10.6	9.71
4	9.4	8.84
5	11.0	9.33
6	10.8	10.16
7	11.3	10.8
8	10.3	8.94
9	10.3	9.66
10	10.8	9.43

Calculate:

(a) The overall mean weight of the nestlings from the two populations.

(b) The variance in weight of the nestlings from the two populations.

(c) Which population has the larger standard deviation?

3. When studying the heritability of a quantitative trait, we commonly analyse its expression in the F_1 and F_2 generations obtained after crossing two highly inbred strains. Which set of progeny provides data for an estimation of the genetic variance?

4. A project was launched to investigate the number of leaf blades produced by different varieties of maize prior to the flower spike. In a cross between two commonly cultivated inbred varieties, the resulting plants showed a variance in leaf number prior to spike production of 1.72 in the F_1 generation and of 4.96 in the F_2 generation.

(a) What is the broad sense heritability in leaf number?

(b) If a farmer wanted to breed a new variety that produced less leaf blades before flowering, does this heritability value suggest a selective breeding programme would be effective?

5. Two different varieties of mangoes produce fruit with the same mean weight of 430 g. One variety shows a low variance for fruit weight, while the other possesses a much higher variance.

   (a) What are the reasons for the differences in variance?

   (b) If you were a commercial mango grower which variety would you grow, and why?

   (c) If you wanted to develop a variety that produced heavier fruit, would you choose the variety with the low or high variance? Explain your choice.

6. Values of the time it took 500 schoolchildren to find their way to the centre of a maze showed an approximately normal distribution, with a mean of 10 min and a standard deviation of 3.2 min. What proportion of the group took

   (a) longer than 16.4 min

   (b) above 6.8 min

   (c) less than 6.8 min?

7. Selective breeding is a common strategy for altering quantitative traits. Suppose a plant breeder started with a wild population of blueberries and subjected the plants to many generations of selective breeding, with the goal of obtaining plants bearing larger berries. How would you expect the heritability of this trait in his cultivated populations to change from that in the original wild population after many generations of selective breeding?

8. Assume three key genes determine the weight of pumpkin fruit: A/a, B/b and C/c. A pumpkin of weight 3 kg and genotype **aabbcc** was crossed with a pumpkin of 6 kg and genotype **AABBCC**. If the presence of each dominant allele results in the addition of an extra 500 g to the developing fruit compared to the recessive homozygote, what weights of pumpkins would you expect in the:

   (a) F_1 generation

   (b) F_2 generation?

9. The following variances were calculated for leaf width in a population of cowslips beside a busy motorway:

   Additive genetic variance = 4.2; dominant variance = 1.6; interactive variance = 0.3; environmental variance = 2.7

   Calculate the:

   (a) broad sense heritability of this trait

   (b) narrow sense heritability of this trait.

10. A salmon breeder wants to increase the rate of growth of his stock. He therefore chooses to breed from the fish achieving the greatest length by 8 weeks. The mean length was 13 cm and so he decided to breed from individuals that had reached 18 cm or more by 8 weeks. If the mean length of salmon by 8 weeks in the next generation was 15.5 cm:

    (a) Estimate the narrow sense heritability for this trait.

    (b) Advise the breeder of the feasibility of his plan to increase the size of his salmon by selective breeding.

11. A farmer has recently acquired a herd of boar and wishes to increase the average body weight. Her current herd has a mean body weight of 325 kg. The farmer decides to breed only from her six heaviest animals who have a mean weight of 350 kg. They produce 39 offspring with a mean weight of 333 kg. Does this indicate a heritable component to body mass among the herd and, so, would you advise the farmer to continue selective breeding to achieve her goal of increasing the average body weight of the herd?

10 An introduction to population genetics

The focus of population genetics is obviously upon populations. This contrasts with the main concerns of Mendelian transmission genetics, described in earlier chapters and, to some extent, the quantitative genetics of Chapter 9, where the focus is upon the genotype of individuals and the distribution of genotypes resulting from a single mating. Population geneticists investigate the patterns of genetic variation shown by individuals within different groups, or populations. They are interested in questions such as:

- How much genetic variation exists in a given population?
- What processes control the amount of observed variation in a population?
- How might this genetic variation change with time?
- What processes are responsible for producing genetic divergence between populations?

Answers to such questions have implications for both evolutionary theory and conservation policy. Indeed the impetus for the development of the discipline of population genetics was the realization, soon after their re-emergence in 1900, of the relevance that Mendel's ideas on inheritance had for Darwinian evolutionary theory. Charles Darwin was aware of the tremendous natural variation shown by individuals within a species. On this observation he had based his theory of **natural selection** – 'that any being, if it varied however slightly in any manner profitable in itself...will have a better chance of surviving... (and) success in leaving progeny'. Darwin was, however, unaware of any underlying mechanism that could account for the morphological (phenotypic) variation he observed. The rediscovery of Mendel's work, 19 years after Darwin's death, and the subsequent development of the concept of the gene and alleles, established an understanding of how a basis of inheritance could lead to variation. During the early 1900s, the British mathematician Godfrey Hardy (1877–1947) and the German physician Wilhelm Weinberg (1862–1937) each independently produced a set of equations enabling the variation present in populations to be expressed in terms of allele and genotype frequencies (Fig. 10.1).

This chapter:

- shows how the variation present in a population can be analysed in terms of allele and genotype frequencies;
- discusses the impact upon the observed variation of changing crucial features of the population such as its size, mating patterns or selective forces.

Fig. 10.1

Godfrey Hardy. Although Hardy's name nowadays is generally associated with the Hardy–Weinberg equation, he was not a geneticist. During the first half of the twentieth century Hardy was one of Britain's leading pure mathematicians, who hoped that his mathematics would never find practical applications! Hardy spent most of his working life at Trinity College, Cambridge. At only 33 years of age he was elected a fellow of the Royal Society, with many other honours following. A private and shy man, he hated having his photograph taken. This snapshot is one of only five known to exist.

10.1 Population genetics: some definitions

A **population** can be defined as a local group of a single species within which mating is actually or potentially occurring. The interbreeding individuals share a common set of genes referred to as the **gene pool**. It is the characteristics of the gene pool that are studied in population genetics; for example, **frequencies** of different alleles and resulting genotypes in the sample group. Remember that a frequency refers to a proportion and always ranges in value between 0 and 1. For example if 19% of people in a group are left-handed, then the frequency of left-handedness in this group is 0.19. The first step in studying the variation present in a population is to measure the frequency of different alleles for individual genes. Population geneticists are often interested in knowing how many loci are **polymorphic**, i.e. possess two or more different alleles with the frequency of the rarest exceeding 0.1%. The more loci that are polymorphic, the more variation the population shows.

10.2 Calculating allele frequencies

Allele frequencies may be calculated in one of two ways. The simplest method is when each of the possible genotypes at a locus produces a separate phenotype. In this case allele frequencies can be calculated directly from the observed numbers of different phenotypes. Such is the case with the autosomally inherited MN blood groups in humans, described below. The second method uses the recessive genotype frequency as a starting point to deduce allele frequencies, as shown in subsequent sections.

Human blood can be classified in many different ways. The M and N blood groups relate to the antigens produced by gene L* on chromosome 2 (Table 10.1). Note that the two blood group alleles are codominant so that each possible genotype produces a different phenotype: also, that the two alleles are generally referred to more simply as M and N.

Table 10.1
Relationship between MN blood group phenotypes and genotypes

Blood group	Genotype
M	$L^M L^M$ or **MM**
MN	$L^M L^N$ or **MN**
NN	$L^N L^N$ or **NN**

Because each genotype produces a different phenotype we can determine the frequencies of the two alleles directly from collected data. Table 10.2 shows the distribution of MN blood groups in a sample of 1000 randomly selected individuals.

Table 10.2
Distribution of MN blood groups in a sample of 1000 individuals

Blood group	Number of individuals
M	360
MN	480
N	160

Box 10.1 shows how the frequencies of the M and N blood group alleles can be calculated for the data in Table 10.2.

The frequency of the different MN blood groups varies enormously among different human populations (Table 10.3). This illustrates how variable in their allele frequencies geographically distinct populations can be. Because there seems to be no strong selective force acting on blood group antigens, the most likely reason for the range of allele frequencies shown by the different populations is chance fluctuation (see Section 10.9). Another plausible explanation could be that the **M** allele confers an advantage on individuals in a cold climate; while it is more favourable in a hot climate to possess **N** alleles!

Box 10.1 Calculating allele frequencies

The frequency of each allele $= \frac{\text{number of copies of an allele}}{\text{total number of alleles}}$

Phenotype	Genotype	Number of individuals	Number of alleles M	N
M	MM	360	360 × 2	
MN	MN	480	480	480
N	NN	160		160 × 2

Total number of alleles = 2000 (each of the 1000 individuals possess two MN blood group alleles)

Frequency of **M** $= \frac{720 + 480}{2000} = \frac{1200}{2000} = 0.6$

Frequency of **N** $= \frac{480 + 320}{2000} = \frac{800}{2000} = 0.4$

Population	M allele frequency	N allele frequency
Australian aborigines	0.178	0.822
Japanese Ainus	0.430	0.570
US blacks	0.532	0.468
US whites	0.540	0.460
US Indians	0.776	0.224
Greenland Eskimos	0.913	0.087

Table 10.3
Frequencies of M and N alleles in different populations

10.3 The Hardy–Weinberg law

In the case of the MN blood groups, because the **M** and **N** alleles are codominant each genotype produces a distinct phenotype. However, if one allele is recessive, then the heterozygote possesses an identical phenotype to the homozygous dominant individual. In such cases, it is impossible to directly determine allele frequencies. Godfrey Hardy and Wilhelm Weinberg each independently developed a mathematical model that enables us to calculate allele frequencies in such cases. Their work is recognized in the **Hardy–Weinberg law**, which is one of the fundamental concepts in population genetics – as important in its context as Mendelian ratios in transmission genetics. The Hardy–Weinberg law gives us a way of determining:

- genotype frequencies if allele frequencies are known;
- allele frequencies if certain genotype frequencies are known.

According to the Hardy–Weinberg law, the three possible genotypes produced by one gene with two alleles will be present in a population in the following proportions:

$$p^2 + 2pq + q^2$$

where: p = frequency of the dominant allele
q = frequency of the recessive allele
p^2 = homozygous dominant genotype frequency
$2pq$ = heterozygote genotype frequency
q^2 = homozygous recessive genotype frequency

Furthermore, if certain conditions exist within a population the genotype frequencies add up to unity, i.e.

$$p^2 + 2pq + q^2 = 1$$

These conditions are that:

- The population under consideration is large, or large enough that sampling error is negligible.
- No genotype possesses a selective advantage or disadvantage, i.e. that all genotypes are equally viable and fertile.
- There is an absence of other factors, such as mutation, migration or genetic drift, which could favour certain genotypes.

If all these conditions hold, then the population is said to be in **genetic** or **Hardy–Weinberg equilibrium**, and frequencies of different alleles and genotypes will not change from one generation to the next.

In developing their useful equation for calculating genotype frequencies within a population, Hardy and Weinberg both applied basic rules of Mendelian inheritance. A summary of the equation's derivation is shown in Box 10.2.

Box 10.2 Derivation of the Hardy–Weinberg equation

Consider a gene with two alleles, **A** and **a**. In population genetics **p** represents the frequency of the dominant allele (**A**) and **q** the frequency of the recessive allele (**a**). Starting with this premise, the Mendelian principles of inheritance that result from the processes of meiosis and sexual reproduction enabled Hardy and Weinberg each independently to deduce that genotype frequencies in any one generation can be expressed as $p^2 + 2pq + q^2 = 1$. A Punnett Square best shows the derivation of this relationship:

Gametes	**p** (A)	**q** (a)
p (A)	$\mathbf{p^2}$ (AA)	**pq** (Aa)
q (a)	**pq** (Aa)	$\mathbf{q^2}$ (aa)

Therefore, with allele frequencies of **p** and **q** among the gametes of each parent, the expected genotype frequencies in the next generation will be $\mathbf{p^2 + 2pq + q^2}$.

10.4 Calculating genotype frequencies

If we know the allele frequencies in a population then the frequencies of the resulting genotypes can easily be determined. Consider a population of rats and the gene for resistance to the poison warfarin. If 70% of the alleles at this locus are the dominant resistant allele (**R**) and the remaining 30% the recessive non-resistant one (**r**), then:

$p = 0.7$
$q = 0.3$

We can now use the Hardy–Weinberg equation, $\mathbf{p^2 + 2pq + q^2 = 1}$, to calculate the distribution of the genotypes expected in this and any subsequent generations, assuming equilibrium conditions exist.

Frequency of the dominant homozygote $= p^2 = (0.7)^2 = 0.49$
Frequency of the heterozygote $= 2pq = 2 \times 0.7 \times 0.3 = 0.42$
Frequency of the recessive homozygote $= q^2 = (0.3)^2 = 0.09$

Equilibrium is confirmed because these calculated frequencies add up to unity:

$0.49 + 0.42 + 0.09 = 1.00$

We rarely, however, initially know allele frequencies in a population. A more realistic situation is to have phenotype data, i.e. to know how many individuals possess the dominant phenotype and how many the recessive one. It is possible, though, to determine allele frequencies from such data, and to then use these to predict genotype frequencies. The starting point in these calculations is to use the frequency of individuals showing a recessive phenotype.

Consider again the Hardy–Weinberg equation, and corresponding phenotypes:

$$\underbrace{p^2 + 2pq}_{\text{dominant phenotypes}} + \underbrace{q^2}_{\text{recessive phenotype}} = 1$$

To calculate genotype frequencies from phenotype data the following procedure can be followed:

- Firstly calculate the frequency of the recessive allele from the recessive phenotype (and also genotype) frequency, q^2

  $q = \sqrt{q^2}$
- Having obtained a value for the recessive allele, it is possible to know the dominant allele frequency, because

  $p + q = 1$

  $p = 1 - q$
- Once values for both the dominant and recessive alleles have been calculated, the Hardy–Weinberg equation can be used to deduce genotype frequencies, as in the previous section.

This method of calculating genotype frequencies is used in a variety of different contexts. Box 10.3 describes a use in medical genetics.

Box 10.3 Determining carrier frequencies

Until recently the Mediterranean island of Sardinia had one of the world's highest incidences of the severe recessive blood disorder, ß thalassaemia. In 1975 1 in every 213 individuals was affected and in some isolated villages the incidence was as high as 1 in 100. As the first stage in a programme aimed at reducing the incidence of the disease, the government wanted to know the level of carriers (or heterozygotes) in the population.

An incidence of 1 in 213 thalassaemia sufferers represents a recessive phenotype frequency of 0.0047. Thus:

$q^2 = 0.0047$

$q = \sqrt{0.0047} = 0.0685$

$p = 1 - q = 0.9315$

Carrier frequency $= 2pq = 2 \times 0.0685 \times 0.9315 = 0.1276$

This means that 12.76%, or 1 in 8, of the population were carrying the allele for ß thalassaemia. In 1977 a public education programme was initiated, using local meetings, posters, pamphlets, radio, television and newspapers, and targeted at couples planning a pregnancy. The intention was to identify carriers and advise prospective parents of their chances of conceiving a child with ß thalassaemia. One member of each couple was tested (a simple blood test could identify carriers). If this first result was positive, the second partner was tested. If both partners were found to be carriers, then they were offered prenatal diagnosis. In the first three years of the programme 694 couples were found to be at risk of conceiving an affected child. As the result of these couples' reproductive decisions, 42 foetuses were diagnosed as possessing the genes for expression of ß thalassaemia, of which 39 were electively aborted. By 1980 the incidence of ß thalassaemia had dropped to 1 in 290. The screening programme has continued and today the incidence stands at 1 in 3600.

Population screening for ß thalassaemia carriers has become widespread in many countries where the disease is prevalent. Similar screening programmes now occur in other populations at risk for other severe recessive disorders; for example the fatal Tay–Sachs disease among Ashkenazi Jews.

10.5 Testing for equilibrium

In describing methods for calculating allele and genotype frequencies it has been assumed that populations are in **genetic equilibrium**, and that the frequencies calculated for one generation will be the same in succeeding generations. It is possible to test whether a population is in genetic equilibrium if heterozygotes can be identified as phenotypically distinct from the homozygous dominant individuals, as is, for example, possible with the MN blood groups. To test for equilibrium a comparison is made between the **observed** genotype frequencies and those **expected** if equilibrium exists. If the frequencies closely match, then equilibrium can be assumed to exist. If, however, there are large numerical discrepancies between observed and expected genotype frequencies, this suggests that a population may not be in equilibrium. A chi-squared statistical test (Section 4.8) should then be performed to determine whether the differences between observed and expected frequencies are significant. Box 10.4 shows the results of a chi-squared test on data obtained for the marsh frog *Rana ridibunda*.

Box 10.4 Testing for equilibrium

Genotypes formed from two alleles (**L** and **M**) of the gene for the enzyme lactate dehydrogenase were studied among 340 individuals of the marsh frog *Rana ridibunda*. The observed numbers of individuals were 133 **LL**, 135 **LM** and 72 **MM**. A chi-squared test was used to assess whether this population was in equilibrium.

Calculation of observed allele frequencies;

Genotypes	No. of individuals	No. of **L** alleles	No. of **M** alleles
LL	133	266	
LM	135	135	135
MM	72		144
		401	**279**

Frequency of **L** allele $= \frac{401}{680} = 0.59$

Frequency of **M** allele $= \frac{279}{680} = 0.41$

From the observed allele frequencies, **expected genotype frequencies** are calculated using the Hardy–Weinberg frequencies of p^2, $2pq$ and q^2:

LL $= p^2$ **LM** $= 2pq$ **MM** $= q^2$

Thus: **LL** $= (0.59)^2 = 0.348$; **LM** $= 2 \times 0.59 \times 0.41 = 0.484$; **MM** $= (0.41)^2 = 0.168$

A χ^2 test is used to assess whether there is any significant differences between the observed and expected frequencies.

Calculating a value for χ^2

Genotype	Observed numbers (*O*)	Expected numbers (*E*)*	*O* – *E*	$(O-E)^2$	$(O-E)^2/E$
LL	133	118	15	225	1.91
LM	135	166	–31	961	5.79
MM	72	57	15	225	3.94
Total	**340**	**340**			**11.67**

*change frequencies to numbers for a χ^2 test

Null hypothesis: there is no difference between observed and expected genotype frequencies.

Significance level = 0.05; degrees of freedom = 2; calculated χ^2 value = 11.67; critical χ^2 value = 5.99.

Because the calculated χ^2 value is greater than the critical χ^2 value, the null hypothesis is rejected. There is, therefore, a significant difference between observed and expected genotype frequencies. The population is **not** in Hardy–Weinberg equilibrium.

10.6 Disturbances of the Hardy–Weinberg equilibrium

The Hardy–Weinberg equation enables us to predict allele and genotype frequencies in populations that are stable and unchanging. It is difficult, however, to find natural populations where all the assumptions required of the Hardy–Weinberg law rigorously exist. Rarely do we find large and randomly mating populations where no mutation, migration or differential reproduction are occurring. Populations are dynamic; changes in size and structure are a natural part of their being. The final part of this chapter briefly considers the various factors that disturb equilibrium and how they might contribute to evolutionary change. We can think in terms of:

1. Factors that produce new variation, i.e. introduce new alleles into a population. **Mutation** is the main means of generating new alleles, although migration plays a subsidiary role.
2. Factors that change the profile of existing variation by altering allele frequencies within a population. **Migration** can have a major influence here along with **genetic drift** and **differential reproduction**.

10.7 Mutation

Within a diploid, sexually reproducing population the gene pool is reshuffled each generation to produce new combinations of genotypes among offspring. These new genotype combinations are the result of the processes of meiosis and the random fusion of gametes. They ensure variability among individuals. These processes do not, however, produce any *new* variation. Only a mutation can create a new allele.

Mutations are changes in genes and occur at random, without regard for any possible benefit or disadvantage to the organism. Thus, when considering the possible impact of mutations on a population it is important to know:

- how frequent a given mutation is;
- the effect of the mutation: is it beneficial, harmful or neutral?
- how fast the new allele spreads among individuals of a population;
- whether the new allele is dominant or recessive in its effect on phenotypic expression.

Mutation rates are generally expressed as the number of new alleles at a given locus per given number of gametes, and are of the order of one new mutation per 100,000 gametes, or 1×10^{-5} gametes. Most new alleles are disadvantageous, reducing the fitness of an individual. It is believed, therefore, that over evolutionary time, mutation rates themselves have been subjected to selection pressures. Thus the generally low mutation rate at most loci represents a balance between the potentially advantageous and disadvantageous effects of producing new alleles. Any one gamete is unlikely to have more than one new mutant allele.

With an average mutation rate of 1×10^{-5} per locus, even if the new allele conferred a selective advantage upon individuals, it would take many thousands of generations for the allele to reach significant levels within a population. Figure 10.2 shows, for example, that it would take in the order of 70,000 generations for a second favourable allele at a previously monomorphic locus to reach an allele frequency of 0.5! The process of mutation is an important evolutionary force within populations in that it is the only means of producing new alleles. It is not, however, a major force by itself in changing allele frequencies as mutation rates are too low. Other factors need to be operating within a population to bring about a significant increase in the frequency of a new allele. Migration and changing selective forces are examples of two such forces.

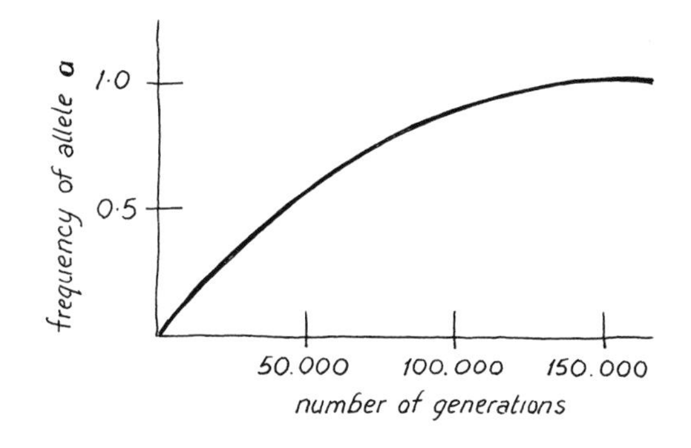

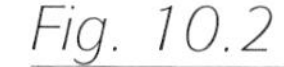
Fig. 10.2

Changes in allele frequency by mutation alone. A population begins with one allele, **A**, at a locus. Allele **A** mutates to **a** at a frequency of 1×10^{-5}. The graph shows the change in frequency of allele **a** with time.

10.8 Migration

Frequently plant and animal species become divided into geographically separate populations. Migration occurs when individuals move between two such populations. If there is a large number of immigrants or a large difference in allele frequencies between the existing population and the immigrants, then a major change in allele frequency can occur in just one generation. Migration can be viewed as both a potentially positive and negative process. The positive aspect is its ability to introduce new alleles, and so novel variation into a population. However, migration abolishes differences between populations. Thus if there were major shifts in selective pressures this could mean no population had a suitable spectrum of variation to respond favourably.

Migration can be a one-off event or a continual process because, for example, two populations are no longer geographically isolated. In the latter case we consider migration more in terms of '**gene flow**'. Much gene flow has occurred, for example, over the last couple of centuries between black and white Americans. Most of the black population of the USA is descended from ancestors in West Africa where allele frequencies at many loci are different to those in the white Europeans who emigrated across the Atlantic. Gene flow, through intermarriage, is gradually reducing the differences at many loci. Annual gene mixing occurs between populations of the monarch butterfly, *Danaus plexippus* (Fig. 10.3).

Habitat destruction is a global conservation issue. It often fragments populations of plant or animal species already under threat of extinction. The result is small isolated populations that often show limited variability. In such cases gene flow is encouraged! Great efforts are directed towards maintaining habitat corridors between fragmented populations to encourage gene flow and, thus, maintenance of different alleles in the populations, so that, in effect, individuals are part of one large population and so avoid the problems of **genetic drift** and **inbreeding** associated with small populations.

Corel 634 (NT)

Fig. 10.3

An overwintering clump of monarch butterflies. The butterfly overwinters in central Mexico. In the spring they begin to fly north towards Canada, mating and egg laying en route. The new adults join the northwards flight. As a result of this great communal migration there is little genetic differentiation between different populations.

Related to a bottleneck is the **founder effect**, when a small group of individuals separate from the main large population and establish a colony in a new location. Again numbers generally grow, but initially drift will be a major factor operating to change allele frequencies and reduce variation. Thus, in both bottleneck and founder situations, the group of surviving or founder individuals may possess different allele frequencies to the original population; for example, a higher incidence of a particular recessive allele which is then passed to many descendants. This situation can explain the above average frequencies of certain genetic diseases observed among isolated human populations or groups who have emigrated to new lands (Table 10.4).

Table 10.4

The founder effect: examples of elevated frequencies of genetic diseases in particular populations

Group	Disorder	Frequency in founder group	Average world frequency
Afrikaans	Familial hypercholesteremia	1 in 30,000	1 in 1,000,000
Ashkenazi Jews	Tay–Sachs disease	1 in 3600	1 in 350,000
Bantu South Africans	Porphyria	1 in 250	1 in 25,000
Hopi Indians	Albinism	1 in 230	1 in 15,000

10.10 Selection

Genetic drift, mutation, migration and selection all bring about changes in allele, and thus genotype, frequencies in populations. However, the changes caused by the first three factors are random. By contrast, the changes brought about under the influence of selection have a directional, or differential, aspect. As the result of selection, alleles favourable to the survival, and therefore reproduction, of an individual will increase in frequency in a population at the expense of alternatives.

In most populations, at any given moment in time, there are individuals with different alleles, and therefore different genotypes. Because of these inherent genetic differences, some of the individuals in a population will possess phenotypes that better adapt them to the prevailing environmental conditions. Such individuals have a **selective advantage** and will survive and reproduce at the expense of others. Thus, certain genotypes, and therefore alleles, are being favoured and, over time, allele frequencies will change in a population. If it is the conscious choice of humans that individuals survive, then we recognize the process as **artificial selection**; for example, when agriculturists choose to breed from those individuals in a crop that show the most resistance to a certain pesticide or the selective breeding of domestic dogs over the past eight millennia (Fig. 10.7). If selection occurs as a result of natural circumstances then it is called **natural selection**.

Corel 766 (NT); Photodisc 50 (NT)

Fig. 10.7

The selective breeding of the domestic dog. Today there are over 100 different breeds of dog, all descended from a wild wolf-like ancestor.

Natural selection can be a major force in changing allele frequencies. In population genetics we attempt to quantify the strength of selection and to calculate its effect on allele and genotype frequencies, and thus upon the variation shown by populations. Two key concepts used by population geneticists are **relative fitness (w)** and the **selection coefficient (s)**. Basically, the first term is an indicator of the relative advantage of possessing certain genotypes, while the selection coefficient represents the downside of a given genotype! Thus for a genotype to possess a fitness of 1 is good news, but if we are told $s = 1$ for a genotype, there is little hope of an individual possessing that genotype surviving long enough to reproduce.

There are many questions that we can investigate if we know allele frequencies (i.e. p and q) and either s or w. At first sight the numerous mathematical models and formulas may seem daunting, but interesting insights into future population structure can be gained from their use. One issue is investigated in Fig. 10.8. Suppose a recessive homozygote was lethal, $s = 1$. We might be interested to know how quickly the recessive allele would disappear from a population. The graph shows that initially there is a rapid drop in frequency of the recessive allele, as selection operates against the recessive homozygotes. Its frequency is halved after two generations, and again after a further four. But after that its rate of reduction declines. This is because many recessive alleles remain in the heterozygote, hidden to the force of selection that operates against phenotypes, and not alleles. A recessive allele only expresses itself in the phenotype of an individual as the recessive homozygote. Because

Fig. 10.8
Changes in allele frequency under selection.

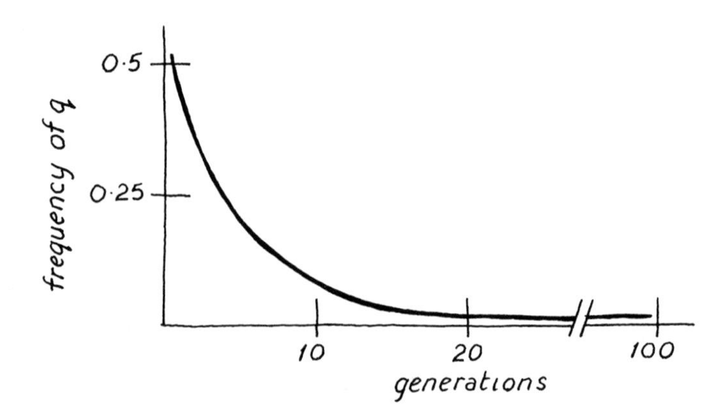

recessive alleles can be present in a population in the heterozygote members, it can be very difficult, indeed impossible, to completely eliminate alleles from a population – which is probably a good thing, in that it maintains variation in a population for 'a rainy day'. This notion can be illustrated by the following classic example of selection in a natural population.

For any starting frequency for the recessive allele the following formula enables us to calculate its frequency (q_n) after a predetermined number of generations, and when $s = 1$ for the recessive homozygote:

$$q_n = \frac{q_o}{1 + nq_o}$$

n = number of generations since original
q_0 = original frequency of q
q_n = frequency of q after n generations

The table and graph show the changes in recessive allele frequency over generations when the initial frequencies of both p and q are 0.5.

Generation	0	1	2	3	4	5	6	10	20	40	70	100
p	0.50	0.67	0.75	0.80	0.83	0.86	0.88	0.91	0.95	0.98	0.99	0.99
q	0.50	0.33	0.25	0.20	0.17	0.14	0.12	0.09	0.05	0.02	0.01	0.01

Prior to the mid-nineteenth century 99% of the population of the peppered moth, *Biston betularia*, was light coloured and, consequently, well camouflaged when resting on the pale lichen-covered trees and buildings. However, as toxic gases produced by factories in the rapidly expanding industrial towns and cities of Victorian England killed the lichens and soot deposits darkened the buildings, the light-coloured moth became an easy target for birds. The rare dark, or melanic, form of the moth suddenly gained a huge selective advantage as it was now the camouflaged variant. By 1900 it formed 95% of populations in industrial centres, although it remained rare in rural areas (Fig. 10.9). Following the passing of laws in the 1950s and 1960s to restrict environmental pollution, lichens have recolonized formerly polluted areas and buildings are gradually being cleaned. The result is a steady decline in the melanic form, as the moth's habitat becomes pale coloured once again. Similar examples of rapid changes in allele frequencies, and so resulting genotypes and phenotypes, include the rapid development of resistance to pesticides by insects and tolerance by grasses to high levels of heavy metals in the soil around mine workings.

Fig. 10.9

The fluctuating selective advantages of the light and dark forms of the peppered moth (*Biston betularia*) at rest on a tree in an unpolluted area.

Summary

- The goal of population genetics is to understand the genetic structure of a population and the forces that determine and change its composition.
- The existence of various alleles at different loci produces genetic variation both between individuals within a population and between different populations.
- A fundamental measurement used in population genetics is the frequency with which different alleles occur at any given locus. For a gene with two alleles their frequency is represented by p and q, and $p + q = 1$.
- Knowing allele frequencies enables us to predict genotype frequencies in a population using the terms p^2, $2pq$ and q^2.
- The frequency of any allele in a population can be changed by recurrent mutation, migration, selection or genetic drift.
- Natural selection is the most powerful force acting on a population's genetic composition, although in small populations genetic drift can also be an important factor.
- A randomly interbreeding population, in which no forces of change are acting, will show constant genotype frequencies for a given locus in successive generations. Such a population is said to be in genetic equilibrium and $p^2 + 2pq + q^2 = 1$.)

Problems

1. The frequency of a recessive allele in a large randomly mating population is 0.2. What is the frequency of:

   (a) the dominant allele

   (b) the heterozygote?

2. In the gaboon viper one main locus determines the lethality of its venom. Individuals with the genotype **VV** are deadly poisonous; heterozygotes, those with **Vv**, are mildly poisonous, while any individuals homozygous recessive, **vv**, are non-poisonous. In a population of 3000 vipers the frequencies of the **V** and **v** alleles were found to be 0.75 and 0.25 respectively. How many of these vipers are non-poisonous?

3. In a large interbreeding population of Canadian hairless cats 81% of individuals are homozygous for a recessive character. What percentage of individuals in the next generation would you expect to be:

   (a) homozygous recessive

   (b) heterozygous

   (c) homozygous dominant?

4. For each of the following, state whether it is an example of an allele, genotype or phenotype frequency:

   (a) Approximately 1 in 2500 Caucasians are born with cystic fibrosis.

   (b) The percentage of carriers of the sickle cell allele in West Africa is approximately 13%.

   (c) The number of new mutations for achondroplasia is approximately 5×10^{-5}.

5. Hurler syndrome is an autosomal recessive disorder of mucopolysaccharide metabolism, resulting in short stature, mental retardation and various bone malformations. The frequency of affected newborn infants is about 1 in 76,000. Assuming random mating, what is the frequency of heterozygotes?

6. A randomly mating population of sheep contains an autosomal recessive allele causing the coat to be curly.

   (a) If the frequency of curly-coated lambs is 10%, what is the frequency of the heterozygous carriers of the allele?

   (b) What is the frequency of heterozygotes among normal fleeced sheep?

7. In certain grasses, the ability to grow in soils contaminated with the toxic metal nickel is determined by a dominant allele. If 78% of the seeds of the meadow grass, *Poa pratensis*, are able to germinate in contaminated soil, what is the frequency of the resistance allele?

8. (a) Name four evolutionary processes that can change allele frequencies in a population.

   (b) Recessive alleles are often lethal when homozygous and so there is selection in each generation against such alleles. Yet such alleles are rarely completely eliminated from a population. Why?

   (c) How does the frequency of heterozygotes in an inbred population compare with that in a randomly mating population with the same allele frequency?

   (d) What does it mean when we say an allele is fixed?

   (e) Why is mutation a weak force for changing allele frequencies?

   (f) What factors cause genetic drift?

9. In a certain population of Australian banjo frogs, 120 are green, 60 are brownish-green and 20 are brown. The allele for brown is denoted **B** and that for green **G**. The two alleles show incomplete dominance.

   (a) What are the genotype frequencies in the population?

   (b) What are the allele frequencies?

   (c) What are the expected frequencies of the genotypes if the population is at Hardy–Weinberg equilibrium?

   (d) Is the population in Hardy–Weinberg equilibrium?

10. Seventy tiger salamanders from one pond in west Texas were examined for genetic variation at the enzyme locus malate dehydrogenase. Two alleles, **A** and **B**, were identified using gel electrophoresis. Among the salamanders the distribution of genotypes was:

Enzyme genotype	Number of individuals
AA	15
AB	49
BB	6

Is this population in Hardy–Weinberg equilibrium?

11 Heredity at the molecular level

In August 1953 the scientific community was thrown into a fever of excitement by a short letter that appeared in the leading scientific journal *Nature*. The authors of the letter proudly announced the discovery of the structure of the molecule of inheritance – DNA. Ten years later the authors would receive the Nobel prize for their insights. To understand the structure of DNA had been the goal of many biologists for much of the previous decade – certainly since 1944, when Oswald Avery (1877–1955) and his co-workers had convincingly shown that DNA was the critical component of the bacterium *Streptococcus pneumoniae* capable of transforming harmless avirulent varieties into pneumonia-provoking versions. From that moment the race was on to elucidate DNA's molecular structure. Francis Crick (b. 1916) and James Watson (b. 1928) reached the winning line first (Fig. 11.1). The model they proposed was supported by all the available evidence and, as they were quick to appreciate, suggested a mechanism whereby the DNA molecule could be perfectly copied: a crucial requirement of hereditary biological material.

This chapter describes:

- the basic structure of DNA;
- the mechanism of DNA replication.

11.1 The structure of DNA – a summary

DNA is a **polymer**, which means, biochemically speaking, it is a huge chain-like molecule whose great size is a result of endless repetition of a much smaller molecule or **monomer**. The repeating monomer of DNA is the **nucleotide**, millions of which join end to end to produce long **polynucleotide** chains. Two of these chains spiral around each other to produce the famous 'double helical' form of DNA.

Fig. 11.1
James Watson (left) and Francis Crick with their model of DNA.

11.2 The biochemical nature of nucleotides

Each nucleotide possesses **three** components:

- a **phosphate group**
- a sugar – **deoxyribose**
- a **nitrogenous base**

Deoxyribose belongs to a group of sugars known as **pentoses** because each sugar contains five carbon atoms. Pentoses exist either as straight chains or pentagon rings. The ring form is found in a nucleotide (Fig. 11.2). Only one carbon atom is shown in the diagram; the other four are at the angles of the pentagon. Furthermore, starting with the right-hand carbon, each is numbered 1′ to 5′ (the numbers and the dash, or 'prime', aid in communicating biochemical structure and function). The nucleotides of DNA are, strictly speaking, called **deoxyribonucleotides**, because of the sugar they contain. However, this cumbersome term is generally simplified to **nucleotide**; although because of the presence of deoxyribose the polymer of nucleotides we are describing is formally known as **deoxyribose nucleic acid**, abbreviated to **DNA**.

The **nitrogenous bases** (so called because they contain nitrogen and combine with hydrogen ions in acidic solution) vary within a nucleotide. There are four main variants to be found in DNA: **adenine**, **guanine**, **cytosine** and **thymine** (commonly represented by their first letters: A, G, C and T). As Fig. 11.2 shows, the four bases differ in their complexity. Cytosine and thymine each possess a single carbon–nitrogen ring and are called **pyrimidines**, while adenine and guanine possess a double ring and are referred to as **purines**.

Within each nucleotide the base and phosphoric acid are bonded to the sugar. The phosphoric acid links to the 5' carbon and the base to the 1' carbon (Fig. 11.2).

11.3 The double helix

Crick and Watson's key contribution towards understanding the nature of DNA was to show how the nucleotides are organized within the molecule. Analysis of all currently available evidence led them to propose the **double helix** model, in which DNA is envisaged as a twisted ladder with bonded bases as its rungs (Fig. 11.3). Alternating sugars and phosphate groups constitute the two sides of the ladder, referred to as the **sugar-phosphate backbone**. The phosphate of one nucleotide links to the 3′ carbon of the sugar of the adjacent nucleotide by a **phosphodiester** bond. At regular intervals the bases project from this sugar-phosphate backbone at an angle of approximately 90° (Fig. 11.3).

Cross-chain binding occurs between these projecting bases and is highly specific. A purine can only bind with a pyrimidine. In fact adenine must link with thymine and guanine with cytosine: the famous **A–T** and **C–G** pairs of DNA. This specific, complementary pairing is the key to understanding much about the structure of DNA, its replication and accessing the stored hereditary information for use by the cell. Regarding the structure of the DNA molecule, it is only the pairing of A with T, and C with G, which orientates the bases in such a way as to produce:

- Weak attractive forces between opposing bases that can hold the two polynucleotide chains together. These forces are known as **hydrogen bonds**.
- A regularly structured molecule; another characteristic feature of the DNA molecule, which was very clearly indicated in the X-ray

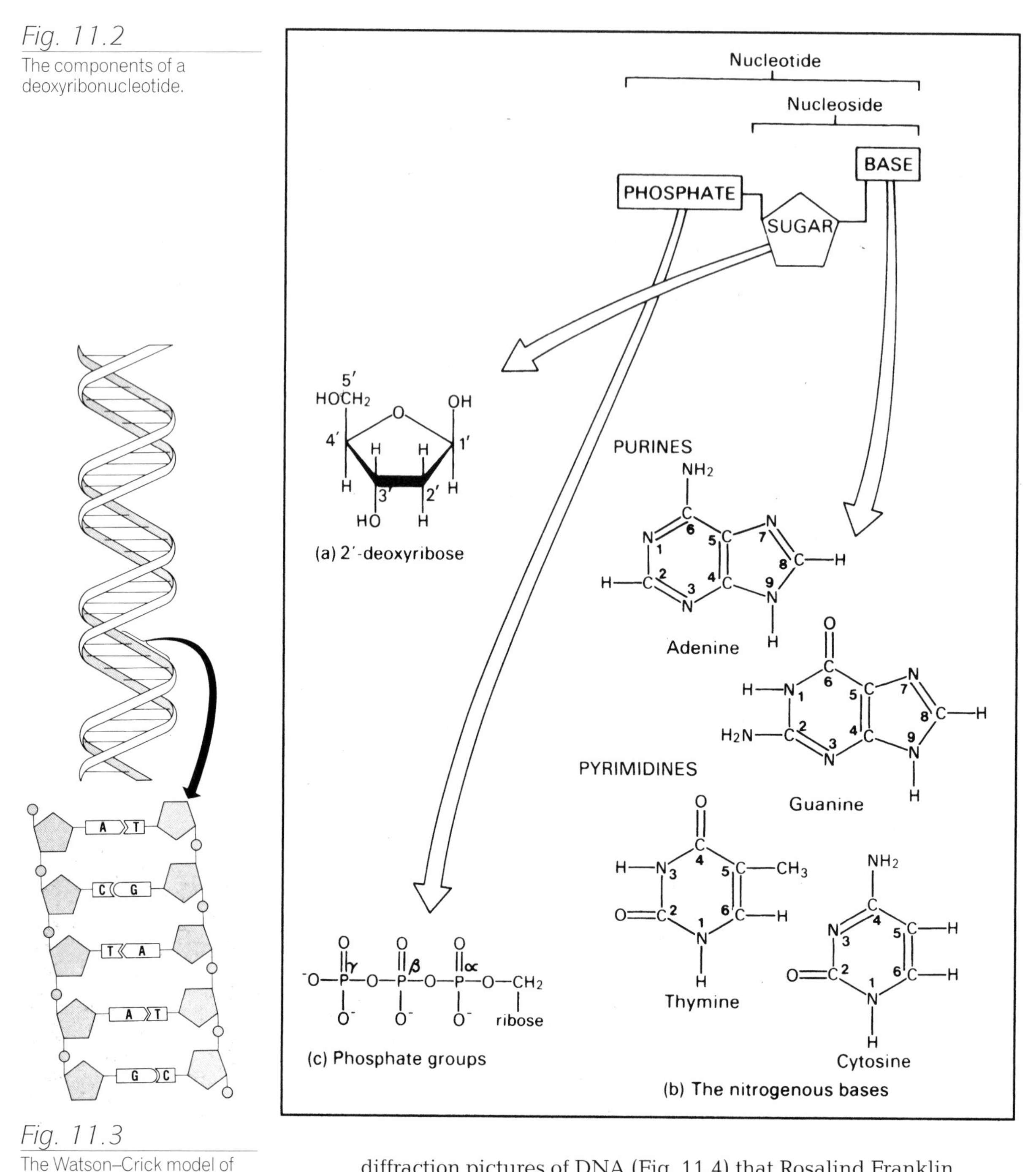

Fig. 11.2
The components of a deoxyribonucleotide.

Fig. 11.3
The Watson–Crick model of DNA. The top diagram is a simplified representation of DNA as a double helix. Below is a short length of DNA untwisted to show the relative positions of the deoxyribose (the pentagons), phosphate and bases. Note also the antiparallel nature of the two strands.

diffraction pictures of DNA (Fig. 11.4) that Rosalind Franklin (1920–58) produced in the early 1950s (Fig. 11.5).

The correct orientation for hydrogen bonding only occurs when the two polynucleotide chains are orientated in opposite directions, i.e. are **antiparallel** (Fig. 11.3).

Each base pair is slightly offset relative to its adjacent ones. This explains the twist or spiral of a DNA molecule – why the polynucleotide chains assume a **double helical** form. One complete twist of the molecule occurs every ten nucleotide pairs, producing predictable and regular dimensions to a DNA molecule. There are just 3.4 Å between adjacent bases; one complete spiral occurs every 34 Å and the width of a DNA molecule is a uniform 20 Å (Fig. 11.6).

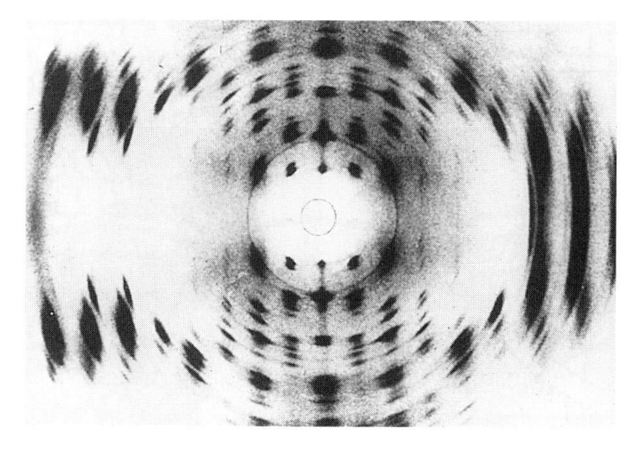

Fig. 11.4

X-ray diffraction pattern of DNA. The technique involves firing a beam of X-rays at a crystal. The atoms in the DNA deflect the X-rays. The scattered X-rays are caught on a photographic plate, producing a pattern that can be used to deduce the arrangement of atoms in the molecule.

11.4 DNA coiling

When describing the structure of DNA we tend to depict the double helix continuing in a long straight line. If, in reality, the approximately three billion nucleotide pairs of the DNA in each human cell was stretched out in this way, it would extend for 2 metres. Yet, this 2 metres of DNA is packaged into a tiny cell nucleus of approximately 10 μm diameter. This packaging is possible because an organized system of coiling the DNA occurs, which reduces its linear length 10,000-fold.

To achieve this condensation the DNA first wraps around groups of **histone** proteins known as **nucleosomes** (Fig. 11.7). Approximately 150 nucleotides are wound around each histone core, while 50 nucleotides form a spacer region before the next nucleosome. The result is a beads-on-a-string effect (Fig. 11.8). A nucleosome string is further folded and compacted into a thicker fibre, sometimes called a **solenoid**. This and the previous, more extended nucleosome arrangement represent the level of organization of the DNA during normal cell functioning, referred to as **chromatin**. To achieve the additional size reduction needed to produce the chromosomes visible during mitosis and meiosis, the chromatin fibres fold again, forming supercoiled loops radiating from a protein core (Fig. 11.9).

Fig. 11.5

Rosalind Franklin. She worked with Maurice Wilkins at Kings College, London and was an expert X-ray crystallographer. Her photographs were vital to Crick and Watson in helping them correctly deduce the structure of DNA. Tragically she died of cancer in 1958 at the age of 37.

new complementary strand could be assembled against each parental template, with the result that two molecules identical to the original one are produced (Fig. 11.10). It took scientists a dozen or so years after the publication of Crick and Watson's model to confirm their mechanism of DNA replication and to work out the molecular details. Matthew Meselsohn and Frank Stahl's elegant experiments were crucial in confirming the overall **semi-conservative** pattern of replication (Box 11.1). Arthur Kornberg (b. 1918) and his coworkers worked out the key molecular details (Section 11.6).

11.6 *Molecular details of DNA replication*

A large number of enzymes and other proteins are employed to ensure accurate DNA replication. For example, before replication can commence the double helix must first be unwound, followed by separation or 'unzipping' of the two strands. **Topoisomerases** are responsible for unwinding the double helix. They achieve this by causing transient breakages in the polynucleotide chains; either in one (type I enzymes) or both strands (type II enzymes). **Helicases** break the attraction between the bases on opposite strands of the parental DNA molecule. **Binding proteins** coat the separated single strands to prevent them re-annealing.

Once single template strands are exposed, **DNA polymerases** align the correct nucleotides opposite their complementary bases and catalyse the formation of new phosphodiester bonds. This complementary copying occurs in a different way on the two template strands because DNA polymerases can only synthesize DNA in a 5' to 3' direction. This means the template must be read in a 3' to 5' direction. But the two strands of DNA are orientated in an antiparallel fashion. Figure 11.11 illustrates the solution to this problem. Only one strand can be synthesized continuously – that copied from the **leading strand**, which is read in a 3' to 5' direction. In order for the DNA polymerase to be able to synthesize in a 5' to 3' direction against the other template strand, replication proceeds in the apparently wrong direction in small sections. The resulting **Okazaki** fragments (named after the biologist who first demonstrated their existence), are eventually joined together by another enzyme, **DNA ligase** (Fig. 11.11). As the new DNA molecules grow they wind around each other to form double helices.

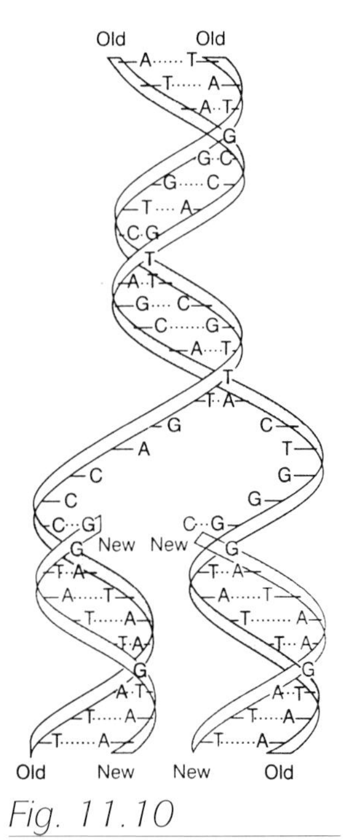

Fig. 11.10

DNA replication. A section of the double helix unwinds to produce two separate template strands. By complementary base pairing two new strands are produced.

The process of continuous replication of one parental strand and discontinuous replication of the other continues until the entire DNA molecule has been duplicated – or at least a section. In most organisms DNA replication would be an extraordinarily time-consuming process if it proceeded linearly from one end of a chromosome to the other. For example, it would take two months for an average-sized human chromosome to be replicated! Instead, replication begins simultaneously at many different sites along a chromosome, called **replication origins** (Fig. 11.12). Eventually the many separately replicated sections join together.

In addition to adding new nucleotides, most DNA polymerases also **proofread** the newly synthesized DNA. If an incorrect nucleotide has been

Box 11.1 Semi-conservative DNA replication

The term 'semi-conservative' refers to the fact that any newly produced DNA molecule contains one original parental strand and one newly synthesized daughter strand. Although Crick and Watson's semi-conservative scheme seemed the most likely, other patterns were suggested in the 1950s, as illustrated below.

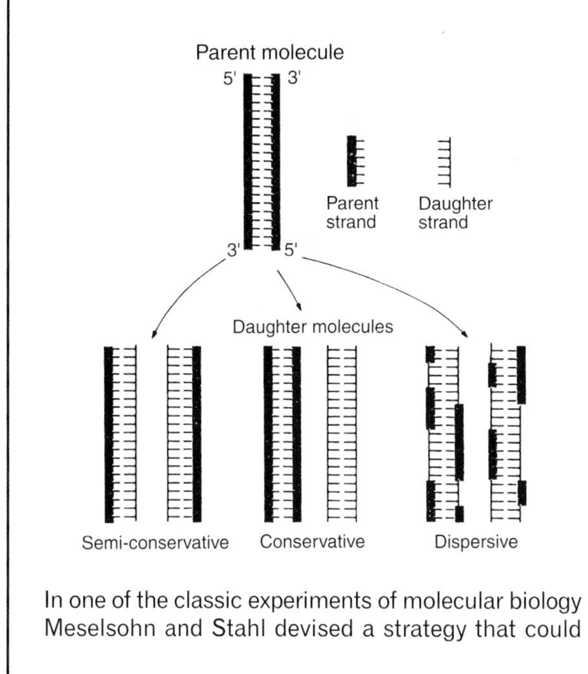

In one of the classic experiments of molecular biology Meselsohn and Stahl devised a strategy that could distinguish between the three proposed mechanisms. They differentially labelled parental and daughter chains. To do this they used different atomic forms or **isotopes** of nitrogen; the normal ^{14}N type and the heavier ^{15}N form. It is possible to control which isotope the nitrogen-containing bases of DNA possess by limiting which form is available in an experimental organism's culture medium. DNA molecules containing one or other or a mixture of the two isotopes will be of slightly different mass and can be separated and identified by **density gradient centrifugation**. Centrifugation separates cellular constituents according to their mass. If DNA is centrifuged through a test tube containing caesium chloride, the latter forms a density gradient and the different DNA molecules sediment or 'band' at different, characteristic positions, according to their mass.

In their experiment Meselsohn and Stahl used the bacterium *Escherichia coli*. Their results were entirely in agreement with a semi-conservative mode of DNA replication. Initially *E. coli* were grown in a medium containing only ^{15}N. At the beginning of the experiment the bacteria were transferred to ^{14}N medium. After one generation the DNA molecules sedimented at a site intermediate between pure ^{15}N and pure ^{14}N. This result eliminated the conservative model, but could support either of the other two. The result, an intermediate band and one at the ^{14}N position, after a further round could only support semi-conservative.

inserted these enzymes can recognize the mismatched pair, excise the wrong nucleotide and replace it with the correct one. Hence DNA replication is a virtually error-free process, e.g. DNA polymerase III of *E. coli* is believed to leave an uncorrected error just once in 5,000,000,000 nucleotides!

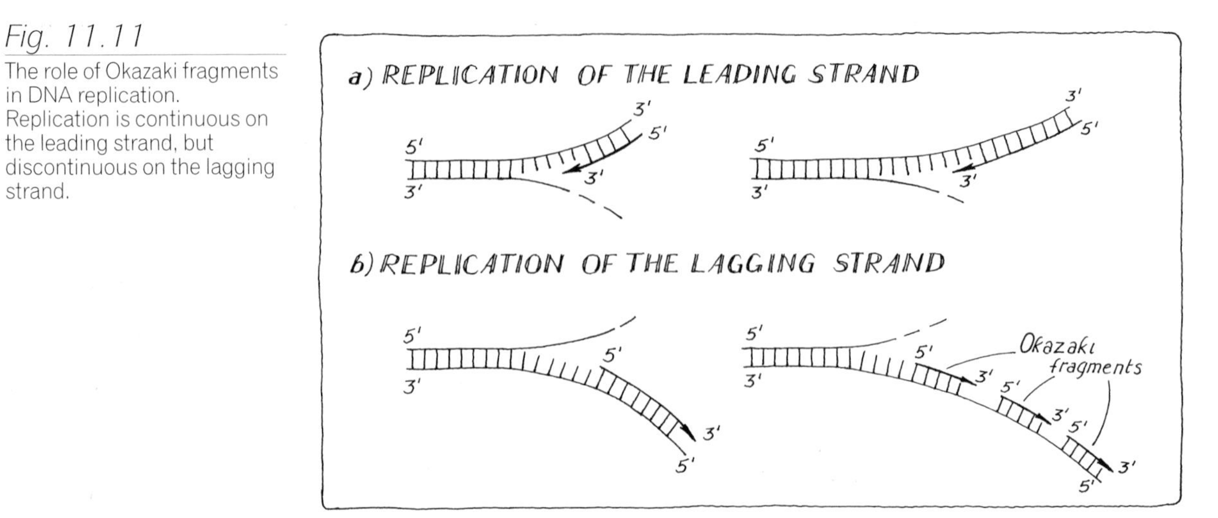

Fig. 11.11

The role of Okazaki fragments in DNA replication. Replication is continuous on the leading strand, but discontinuous on the lagging strand.

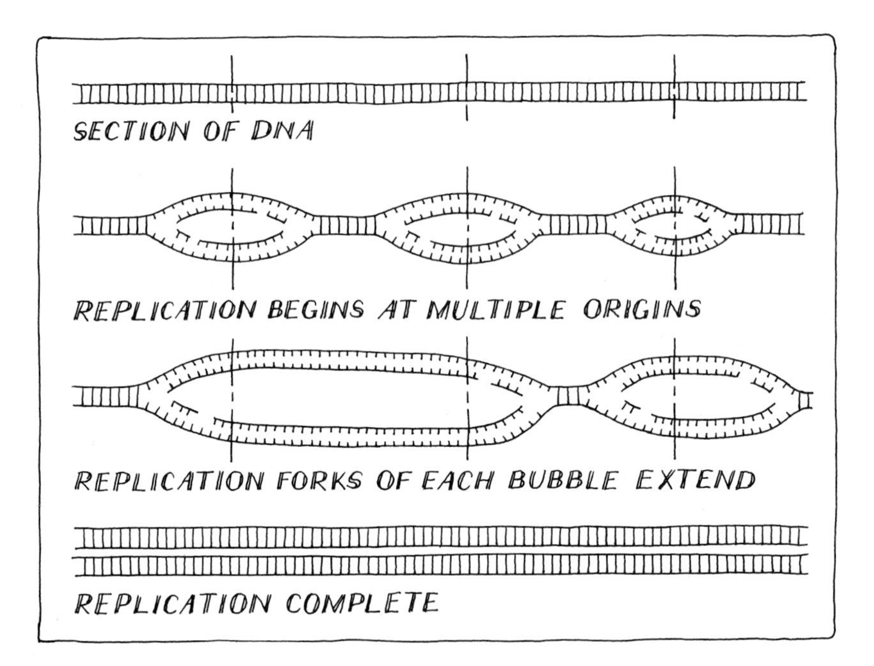

Fig. 11.12

Replication origins and their movement along a DNA molecule.

Summary

- The genetic material is DNA.
- DNA consists of two antiparallel polynucleotide chains held together in a double helix by hydrogen bonds between the chains.
- Each nucleotide consists of a sugar (deoxyribose), a phosphate and a nitrogenous base.
- Purines and pyrimidines are the two types of bases. Adenine and guanine are purines, while cytosine and thymine are pyrimidines.
- Complementary pairing of bases occurs across chains: adenine with thymine and guanine with cytosine.
- Exact replication of a DNA molecule occurs in a semi-conservative fashion. Each separated parental strand acts as a template for a complementary copy to be synthesized.
- Because of the antiparallel nature of the double helix, DNA is synthesized continuously only on one strand. On the opposite strand, synthesis produces short fragments that are later joined.
- DNA polymerase is the main DNA synthesis enzyme. It can also detect and correct misincorporated errors.

Problems

1. In a fragment of double-stranded DNA there are a total of 200 base pairs of which 45 are thymine. How many of the following are there?

(a) nucleotides (b) complementary base pairs

(c) adenine molecules (d) deoxyribose molecules

(e) cytosine molecules

2. A single strand of DNA contains the base sequence 5'-ACCGGTAGAATCG-3'. A complementary strand is synthesized from this template strand.

(a) What is the sequence of the new strand?

(b) In which direction will the DNA polymerase move along the template strand?

3. If 27% of the bases of an organism's DNA are guanine, what percentage are adenine?

4. In DNA, the proper alignment of bases needed for hydrogen bonding of complementary bases occurs when the two polynucleotide chains are orientated in (a) the same direction, or (b) opposite directions?

5. The double-stranded DNA molecules extracted from a newly discovered virus were found to be 102 μm in length.

(a) How many complete turns of the two chains are present?

(b) How many nucleotide pairs are present in one such molecule?

6. For double-stranded DNA which of the following base ratios are always equal to 1?

(a) $A + T/G + C$ (b) C/G

(c) $T + G/C + A$ (d) A/G

7. Indicate whether each of the following statements are true or false.

(a) $A + T = G + C$

(b) The two strands of a DNA double helix are identical.

(c) If the base sequence on one DNA strand is known, the sequence of the second can be deduced.

(d) The structure of DNA is invariant.

(e) If there are 34% adenine bases in a DNA molecule, then there will also be 34% cytosine bases.

8. What is meant by the terms semi-conservative and conservative replication?

9. Why is DNA continuous alongside one template strand and discontinuous alongside the other?

10. If 39% of the DNA for the slime mould *Dictyostelium discoides* is adenine, what can you conclude about the base composition of this species' DNA?

11. The double-stranded DNA molecule of the Epstein–Barr virus contains 172,280 base pairs.

(a) How many nucleotides are present?

(b) What would be the length of this molecule?

12 From genes to proteins

Why should some tomato plants produce red fruit and other plants yellow ones? The explanation offered in Chapter 2 for these different phenotypes referred to the different fruit colour alleles the two types of plants possessed, but was not particularly concerned with considering the nature of the obviously different information the two types of alleles contained and how that influenced phenotype. The focus of Chapter 2 was, after all, a discussion of inheritance patterns and ways of predicting what colour fruit future generations of plants might produce. This and the previous chapter are, however, concerned with different issues. They address the nature of the information contained within individual genes and how it is accessed and used by a cell. Chapter 11 described the molecular nature of a gene's key constituent, DNA, and how it can be accurately copied each time a new cell is produced. This chapter considers the expression of the information contained within a gene.

The key molecule in this chapter's story is protein (Box 12.1). Most genes contain the information for the production of a specific protein, which directly or indirectly leads to expression of a particular phenotype. Molecular biologists have worked out how the DNA of genes codes for proteins and can describe, in considerable detail, how a cell produces a protein from the encoded information. Much less is understood, however, about the way a given gene's protein results in a final phenotype. This chapter will address these three key areas of gene expression, i.e.

- the nature of the genetic code;
- the flow of information from DNA to protein;
- the expression of different phenotypes.

Box 12.1 The nature of a protein molecule

Proteins are polymers consisting of chains of amino acids linked by peptide bonds. Twenty different amino acids are found within proteins. Each different protein has a different specific amino acid sequence. Because a single protein commonly consists of several hundred, even sometimes thousands, of amino acids, and at each site in the chain there are 20 possibilities, the variety of proteins is virtually limitless!

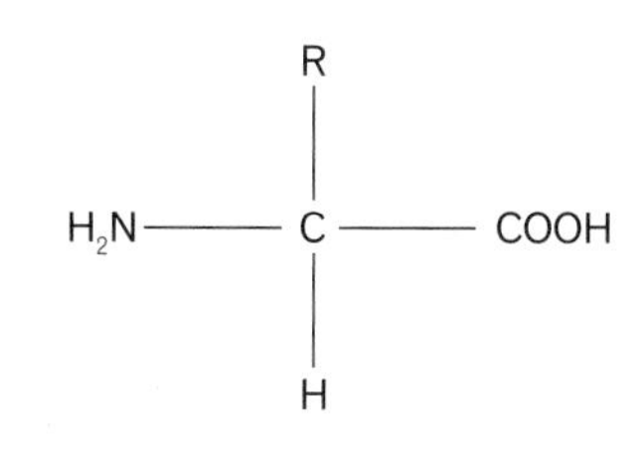

Generalized structure of an amino acid
H = hydrogen atom
COOH = carboxyl (acid) group
NH_2 = amino group
R is a different group for each amino acid

The terms **protein** and **polypeptide** are often used interchangeably, but they in fact refer to different aspects of protein structure. A polypeptide is a single chain of amino acids. The term protein should be reserved for the functional molecule. Often a single polypeptide chain is the functional protein. In such cases protein and polypeptide can be used synonymously. Other proteins, however, consist of two or more polypeptide chains. The two terms, then, refer to different states.

The chain of peptide-bonded amino acids represents the first level, or **primary structure**, of a protein. The chain assumes other specific forms before the polypeptide is functional. The **secondary structure** results when chains spiral to produce an **α helix** or different regions associate forming a **β sheet**. Some proteins, often those with a structural role, are functional at this secondary level. The functions of many other proteins (for example, enzymes) are dependent upon a specific 3D shape that results when chains in their secondary state fold, producing the **tertiary structure**.

Secondary and tertiary structures are stabilized by a variety of interactions between the amino acids, including hydrogen and ionic bonding. As stated earlier, two or more polypeptide chains sometimes associate to produce a functional protein. The multi-subunit form of a protein represents its **quaternary structure**.

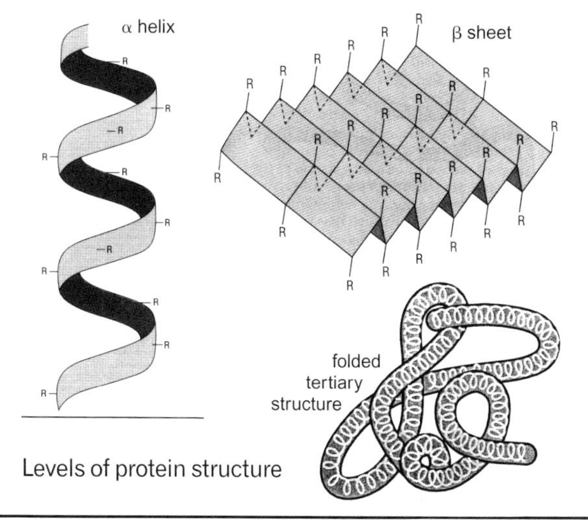

Levels of protein structure

12.1 Information flow from gene to protein – a summary

A gene comprises a linear sequence of nucleotide pairs, while a protein is a linear sequence of amino acids. There is a direct and specific relationship between the two sequences: the nucleotide sequence determines, and correlates with, the amino acid sequence. Each consecutive three nucleotides specifies a single amino acid of the encoded protein. This colinearity of sequences between a gene and a protein is illustrated in Fig. 12.1 and is discussed in more detail in Section 12.6.

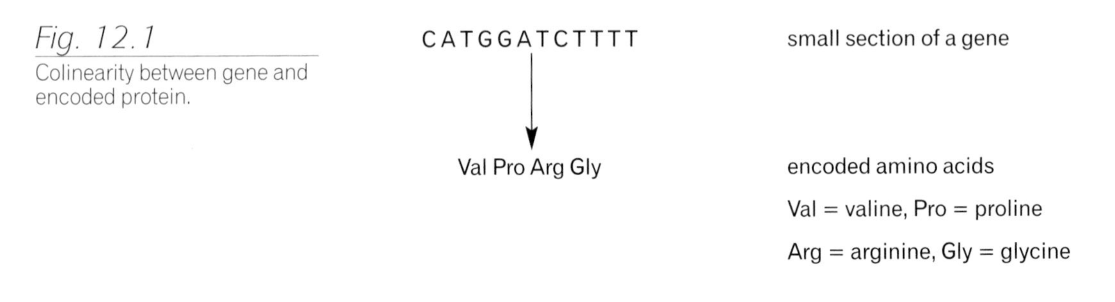

Fig. 12.1
Colinearity between gene and encoded protein.

The encoded information and the site of protein synthesis are, however, physically separated. Synthesis of proteins occurs at ribosomes within the cell cytoplasm, yet the instructions for their synthesis reside within DNA in the cell nucleus. Nucleus and cytoplasm communicate via a mobile information carrier, RNA (ribonucleic acid). Thus there is a flow of information from DNA into a protein molecule. This flow is often referred to as the **central dogma** of molecular biology.

DNA → RNA → protein

The production of RNA from DNA is known as **transcription**. The use of the information within an RNA molecule to manufacture a protein is referred to as **translation**. Because RNA is a pivotal molecule within the processes of transcription and translation, its biochemical nature will first be discussed.

12.2 The biochemical nature of RNA

RNA, like DNA, is a polymer composed of nucleotides. Each ribonucleotide, as does its DNA counterpart, comprises a phosphate group, a pentose sugar and a nitrogenous base. There are, however, two important differences between the nucleotides of RNA and DNA:

1. The pentose sugar is slightly different in the two molecules. In RNA it is ribose, instead of the deoxyribose in DNA. Ribose possesses an extra oxygen atom attached to the third carbon [Fig. 12.2(a)].

2. As in DNA, there are four different nitrogenous bases found within RNA. Three are the same in the two molecules – adenine, guanine and cytosine. However, instead of thymine, RNA contains uracil. Uracil is structurally similar to thymine and can pair with adenine. It possesses a single hydrogen atom, instead of a methyl group (CH_3) on carbon 5 [(Fig. 12.2(b)].

Phosphodiester bonds link adjacent nucleotides in RNA, as in DNA (Section 11.3). However, whilst DNA is double-stranded, RNA usually only occurs as a **single-stranded chain**. Furthermore, because each RNA molecule contains the information relevant to just one gene, it is much shorter in length than a DNA molecule: thousands, rather than millions, of nucleotides.

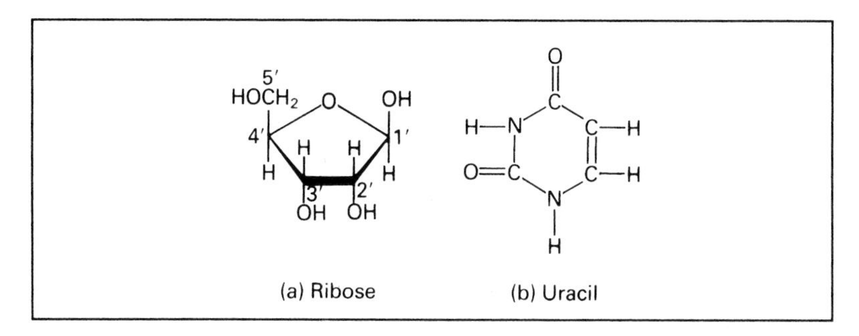

Fig. 12.2
Biochemical structures of (a) ribose, (b) uracil.

12.3 Producing RNA: transcription

Transcription is the process by which an RNA molecule is formed in a similar manner to DNA replication, i.e. a DNA strand is used as a template for synthesis of a complementary one. The process is illustrated diagrammatically in Fig. 12.3. As in DNA replication, the DNA unwinds, hydrogen bonds between complementary base pairs of a gene are broken, and the two polynucleotide strands separate. Now, in contrast to DNA replication, only one of the two DNA strands is a template for RNA synthesis. Free ribonucleotides pair with the exposed bases of the **template** strand and the sugar and phosphate groups bond producing a new single-stranded RNA molecule, which leaves the nucleus via a pore in the envelope. The transcribed RNA strand will, therefore, possess an identical sequence (except for the substitution of uracil for thymine nucleotide) to the DNA strand that did *not* serve as the template (Fig. 12.3).

Because only a short region of the DNA molecule, a gene, is transcribed into an RNA molecule, this raises the question as to how the enzymes know where to begin and end RNA synthesis, or indeed which genes should be transcribed and when. **RNA polymerase** plays a role here. As during DNA replication, a polymerase, this time RNA polymerase, is the principal transcription enzyme. RNA polymerase recognizes and binds to a special start or **promoter** sequence, a short distance in front of, or **upstream** from the gene to be transcribed. Once bound, the polymerase moves along the template

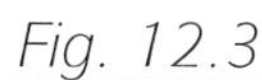

Fig. 12.3
Representation of transcription.

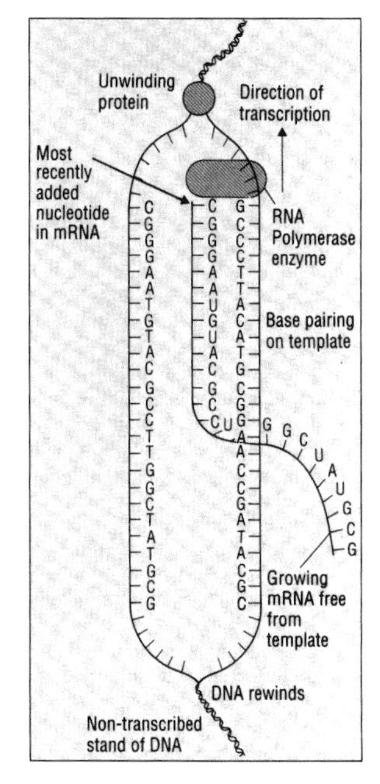

strand in a 5' to 3' direction, catalysing polymerization of the ribonucleotides, until a termination sequence is reached. This signals the end of transcription and also the addition of a polyA tail, i.e. a stretch of up to 250 adenine nucleotides. This, and a special guanine cap at the beginning of the RNA transcript, are believed to protect the new transcript from degradation by cellular nucleases.

12.4 The different forms of RNA

All RNA molecules are produced in the same way, by transcription, from a DNA template and they all possess the same basic structure, as described in Section 12.2. They do not, however, all function in the same way. There are three main types of RNA, each with a different role in the cell. These three types are **messenger RNA (mRNA)**, **transfer RNA (tRNA)** and **ribosomal RNA (rRNA)**.

Messenger RNA is the information carrier. Its nucleotide sequence represents the instructions for assembly of a precise amino acid sequence. It is the intermediary in the flow of information from DNA to protein. With regard to the other two types, the RNA molecule is the final product of gene expression (Fig. 12.4). **Transfer RNA** brings amino acids to the messenger RNA so they can be incorporated into a growing polypeptide chain. **Ribosomal RNA** is a key component of ribosomes, the protein synthesis factory.

Fig. 12.4
The three major types of RNA produced by transcription.

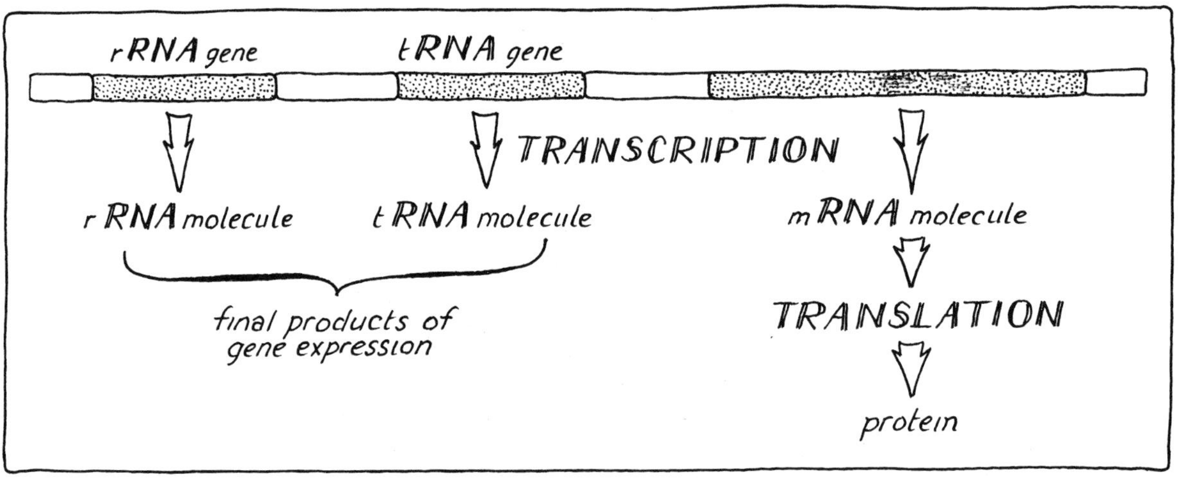

Transcription produces an extended single-stranded molecule. Only mRNA is functional in this state. The transcribed single strands of tRNA and rRNA fold into characteristic three-dimensional forms of which different regions mediate different aspects of translation. Although mRNA is functional in an unfolded single-stranded state, this is only after sections of the original transcript have been removed!

12.5 Exons and introns

In the mid-1970s molecular biologists were surprised to discover that the information within most eukaryotic genes is not continuous, but broken into coding and non-coding regions. The sections containing the biological information became known as **exons**, and the intervening segments as **introns**. The RNA molecule produced by transcription includes both introns and exons. Thus, before mRNA can be used to direct the synthesis of proteins, the introns have to be removed and the remaining coding sections **spliced** together. Specific signal sequences at the boundaries of exons and introns signal splice sites. Splicing is a potentially risky process. If an intron is not precisely removed then quite obviously the information is changed, with the consequence that the encoded protein may not function as well – or at all (Fig. 12.5)!

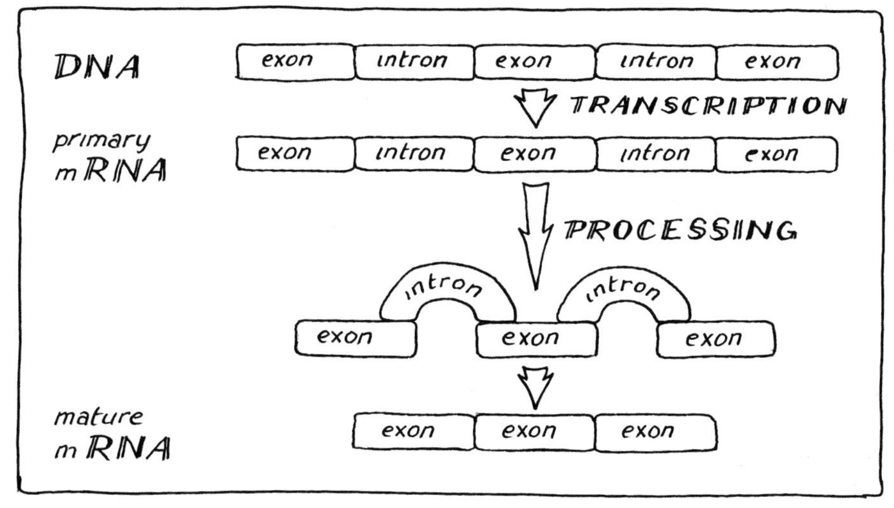

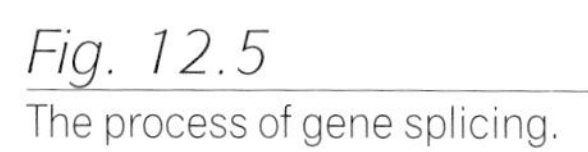
Fig. 12.5
The process of gene splicing.

It is far from clear why the important hereditary information should be broken into chunks, sometimes a great many – the human dystrophin gene (which causes the degenerative muscular disorder, muscular dystrophy, when faulty) has 79 exons. Some interesting examples of ways in which the presence of introns appear to maximize the use of a gene's base sequence have emerged. For example, some genes contain alternative splice sites. These allow the same primary transcript to be spliced in different ways, ultimately producing different protein products from the same transcript. There are several cases of separate genes encoded within the introns of large genes. For example, both the human neurofibromatosis and clotting factor genes contain other, unrelated, genes within their introns!

Other, evolutionary based, explanations have been presented for the existence of split genes. Attention has, for example, been drawn to the fact that, in at least some proteins, each exon codes for a different functional subcomponent. This observation has prompted the suggestion that during evolution exons from different genes can recombine to produce new, advantageous, combinations of biological information. Such a process is potentially more likely to produce new functional proteins than random arrangements of a gene. However, whatever the reason for the evolution of split genes the advantages must outweigh the potential harm that can result from inaccurate splicing.

12.6 The genetic code

In the next section we consider details of the mechanism by which the information encoded within the nucleotide sequence of mRNA is translated into a protein of a specific amino acid sequence. But first, we should consider the nature of the code. This section examines the main properties of the **genetic code**.

The code is a **triplet** or three-letter code. Each consecutive three bases, known as a **codon**, represents an amino acid (Fig. 12.1). There are 64 different ways the four DNA bases can be organized into triplets. Thus, there are 64 codons. Yet there are only 20 different amino acids found in proteins. This means that most amino acids are represented by more than one codon, i.e. the code is **degenerate**. Molecular biologists realized in the early 1960s, when much effort was being devoted to cracking the genetic code, that the code had to be degenerate because a triplet code represented the minimum number of bases that can encode all 20 amino acids. A code based on one base per amino acid gives only four variants, while couplets (e.g. G followed by A, or C and G) can only produce 16 variants. Table 12.1 shows how organisms deal with the excess codons. The different codons for a given amino acid are generally closely related. It is only the last base of the triplet that varies between codons – see, for example, arginine or lysine. Three codons, the **stop** codons, do not represent amino acids. Instead they signal the end of a protein during translation.

Table 12.1

The genetic code, as it applies to mRNA

Codon	Amino acid	Codon	Amino acid	Codon	Amino acid	Codon	Amino acid
UUU	phe	UCU	ser	UAU	tyr	UGU	cys
UUC		UCC		UAC		UGC	
UUA	leu	UCA		UAA	stop	UGA	stop
UUG		UCG		UAG		UGG	trp
CUU	leu	CCU	pro	CAU	his	CGU	arg
CUC		CCC		CAC		CGC	
CUA		CCA		CAA	gln	CGA	
CUG		CCG		CAG		CGG	
AUU	ile	ACU	thr	AAU	asn	AGU	ser
AUC		ACC		AAC		AGC	
AUA		ACA		AAA	lys	AGA	arg
AUG	met	ACG		AAG		AGG	
GUU	val	GCU	ala	GAU	asp	GGU	gly
GUC		GCG		GAC		GGC	
GUA		GCA		GAA	glu	GGA	
GUG		GCG		GAG		GGG	

The abbreviated names of the 20 amino acids are: ala = alanine, arg = arginine, asn = asparagine, cys = cysteine, gln = glutamine, glu = glutamate, gly = glycine, his = histidine, ile = isoleucine, leu = leucine, lys = lysine, met = methionine, phe = phenylalanine, pro = proline, ser = serine, thr = threonine, trp = tryptophan, tyr = tyrosine, val = valine

The code, as outlined in Table 12.1, is universal; i.e. it applies to all living organisms. One exception is mitochondria, which have their own extranuclear DNA (Box 12.2). Several codons encode different amino acids to the same nuclear codes. The general universality of the genetic code represents,

Box 12.2 Extranuclear inheritance

Both mitochondria and chloroplasts contain their own genome. It codes for some of the proteins required in these organelles (for example respiratory proteins in the mitochondria and photosynthetic enzymes in the chloroplasts), and tRNAs, rRNAs, enzymes and proteins involved in translation of these proteins, because mitochondria and chloroplasts have their own translation machinery. Indeed, the general features of these organelles' transcription and translation components are more similar to those in bacterial cells than the eukaryotic cytoplasm in which they are found. Such observations lend support to the **endosymbiotic theory** of the origins of these organelles, i.e. that they were once free-living bacteria that invaded and subsequently established mutually beneficial relationships within proto-eukaryotic cells.

In keeping with a prokaryotic origin for these organelles, their DNA molecules are generally circular and not aggregated with histone or similar proteins. Multiple copies are present; for example, each human mitochondrion has, on average, ten identical DNA molecules. This means that metabolically active cells, with large numbers of mitochondria, possess thousands of copies of the mitochondrial genome. The genomes of these organelles are also much smaller than their nuclear counterparts: for this reason they were among the first DNA molecules to be sequenced. The human mitochondrial genome contains 16,569 base pairs compared to the estimated three billion of the nuclear genome!

The discovery in the 1950s that DNA was present in mitochondria and chloroplasts provided an explanation for some unusual, non-Mendelian, inheritance patterns that had been puzzling geneticists for half a century. Occasionally traits were observed that seemed to be inherited through the female line – for example in 1909, Carl Correns was investigating the inheritance of leaf colour in the four o'clock plant, *Mirabilis japonica*. He was perplexed by the fact that the progeny phenotype was always the same as the female plant. Other examples of traits, seemingly inherited through the female line, continued to be discovered. Now we know that leaf colour in this plant is controlled by a chloroplast gene. Chloroplasts and mitochondria are maternally inherited, because it is the egg that contributes the cytoplasm to the zygote, and therefore the next generation. Any traits controlled by genes within either the mitochondria or chloroplast genome must, therefore, be inherited from the female parent. A few rare human conditions are inherited in this way, for example Leber hereditary optic neuropathy (LHON), the result of a mutation in the mitochondrial NADH dehydrogenase gene. LHON is characterized by rapid loss of vision as the result of optic nerve death. It typically begins in the third decade of life and is usually irreversible. The pedigree below shows the inheritance of this disease within a family.

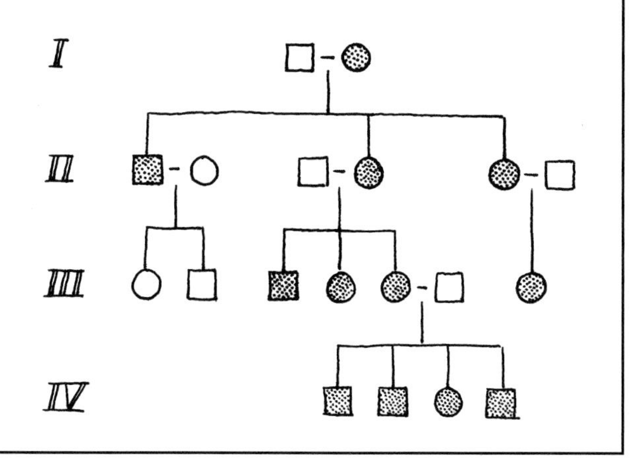

however, persuasive evidence that all organisms have evolved from a common ancestor. Indeed the evolutionary relatedness of organisms is further strengthened by the universal nature of the whole genetic mechanism – the structure of DNA, its mode of replication and the mechanisms of information flow from gene to protein being described in this chapter.

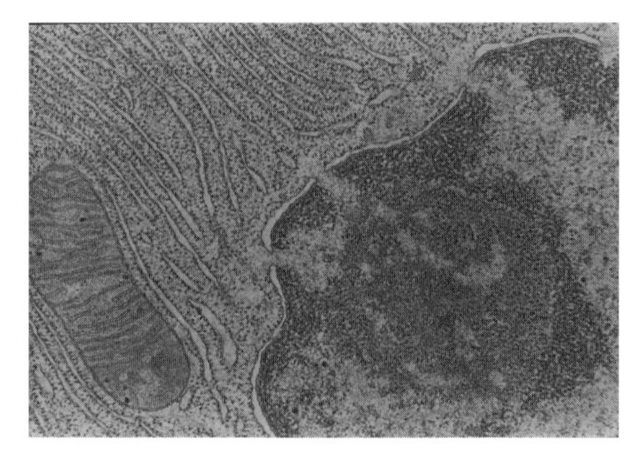

Fig. 12.6

Ribosomes can be seen in this electron micrograph as small black dots covering the surface of the long, thin interconnected tubules of the endoplasmic reticulum.

12.7 Producing the protein: translation

During translation the base sequence of an mRNA molecule is converted into the amino acid sequence of a polypeptide chain. It is a complex process involving the three main types of RNA – mRNA, tRNA and rRNA (Section 12.4) – as well as many proteins. It occurs in the cytoplasm at ribosomes (Fig. 12.6).

Fig. 12.7

Synthesis of a polypeptide by a ribosome.

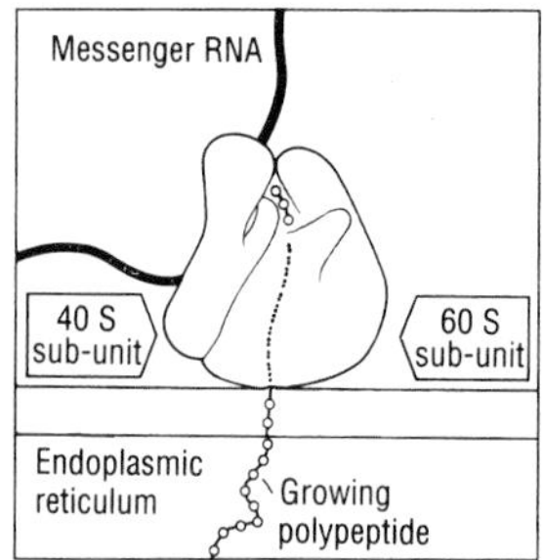

Ribosomes, as indicated in Section 12.4, consist of rRNA and proteins. Some of the proteins function as structural components of the ribosomes. Most have roles to play within translation. Ribosomal RNA itself is of two main types; one type being found in each of the two subunits of a ribosome (Fig. 12.7). During translation the ribosome binds to, and correctly orientates, the mRNA and tRNA molecules for accurate and efficient translation.

Although the mRNA provides the instructions for synthesis of a polypeptide, it cannot bind directly to amino acids. Instead, mRNA interacts with tRNA molecules that carry amino acids and bring them to the mRNA as they are required during translation of the encoded instructions. Each tRNA molecule bonds with and carries one amino acid. Any one tRNA molecule is always loaded with the same type of amino acid. The type is determined by a tRNA's **anticodon**, a specific triplet of nucleotides within the molecule (Fig. 12.8).

Fig. 12.8

Structure of a tRNA molecule. Some nucleotides occupy the same positions in all tRNAs; these are shown in grey. Others vary according to the particular tRNA. The symbols T, D, γ and ψ represent unusual nucleotides found in tRNAs.

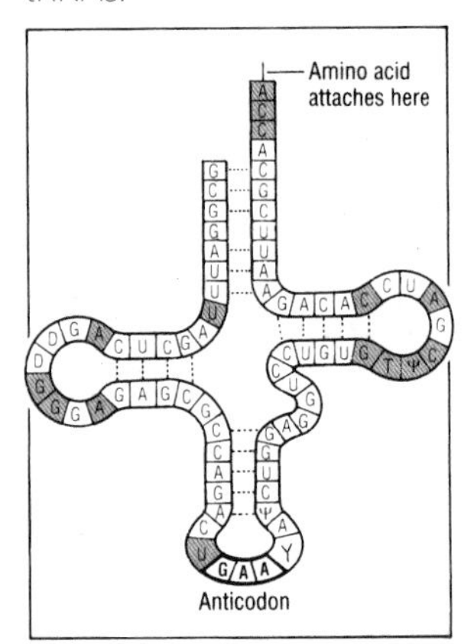

A tRNA molecule generally consists of a sequence of between 80 and 90 ribonucleotides. Segments of the single-stranded transcript are complementary. Thus, the relevant bases pair and so initiate folding of the molecule into a characteristic cloverleaf shape within which there are both double-stranded and single-stranded regions, and two key functional areas – an attachment site for an amino acid and the anticodon (Fig. 12.8). In addition to determining the amino acid that a tRNA molecule carries, the anticodon is also the site of interaction between a tRNA and an mRNA molecule.

Figure 12.9 depicts the process of translation. It starts when the small subunit of a ribosome binds to the initiation site, codon AUG*, near the 5′ end of an mRNA molecule. This binding exposes two mRNA codons to the

*Because the first codon exposed during translation is always UAG, the first amino acid of a new polypeptide is always methionine. It is, however, usually removed before synthesis of the polypeptide is complete. Methionine is also found within polypeptide chains. The ribosome is able to identify the start UAG codon, and know where to initially bind, because there is a nearby tagging sequence of AGGGA.

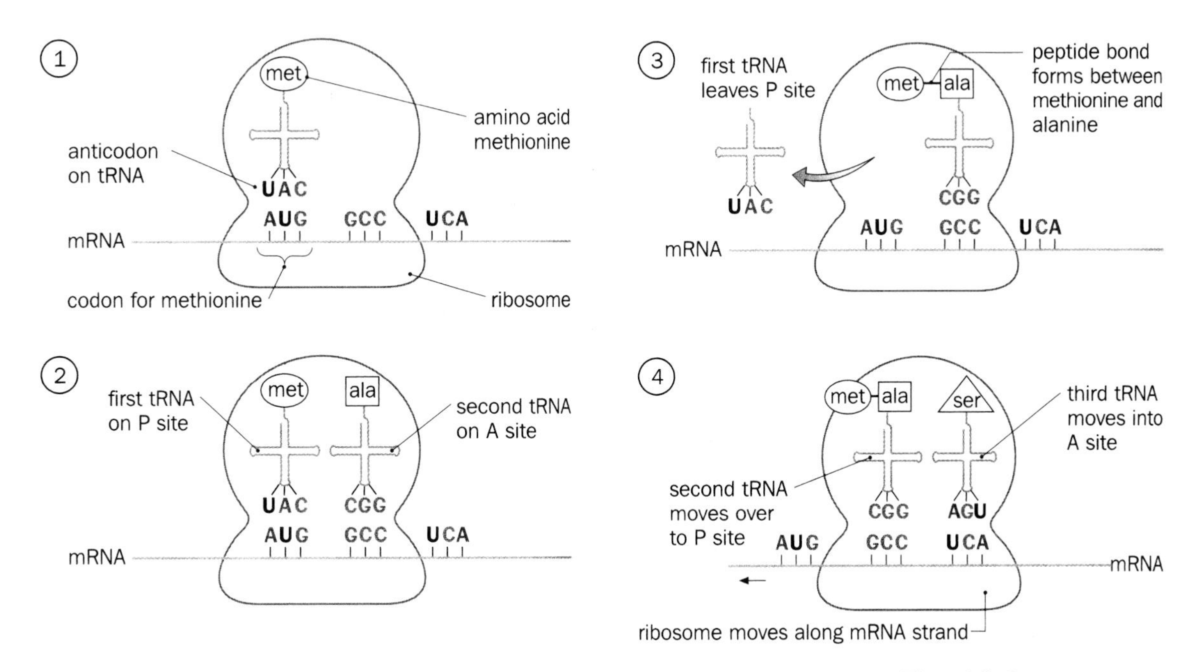

Fig. 12.9
The process of translation.

large ribosome subunit, within which there is room to accommodate a pair of tRNA molecules. Two tRNAs, whose anticodons are complementary to the exposed mRNA codons, now enter the ribosome and hydrogen bond with the mRNA codons. In the example depicted in Fig. 12.9, the two tRNAs must have anticodons UAC and CGG. They will be carrying the amino acids methionine and alanine, respectively. The first peptide bond can now be formed. Using the enzyme peptidyl transferase, methionine, the amino acid from the tRNA in the p site, is transferred and bonded to the amino acid attached to the second tRNA.

Following formation of a peptide bond, the tRNA that lost its amino acid is released into the cytoplasm, where it will quickly pick up a replacement. The ribosome then moves three bases along the mRNA, so that the next codon is within the large subunit and is exposed for translation [(Fig. 12.9(c)]. A third tRNA whose anticodon is complementary to the next mRNA codon now enters the ribosome. A second peptide bond forms, with the growing peptide chain being transferred and linking to the amino acid on the more recently entered tRNA. A peptide of three amino acids has now been produced.

The cycle of tRNA binding, peptide formation and tRNA release continues until a stop codon is reached. This signals to the translation machinery that its job is complete. The mRNA, ribosome and polypeptide separate and the polypeptide assumes its functional conformation. Generally a number of ribosomes bind simultaneously to the same mRNA molecule, each ribosome being at a different stage in the synthesis of the encoded protein. This compound structure is known as a **polysome** (Fig. 12.10) and helps to maximize the amount of a polypeptide synthesized.

Fig. 12.11

SEM of red blood cells from an individual suffering from sickle cell anaemia. Only one cell (top left) is the normal biconcave shape.

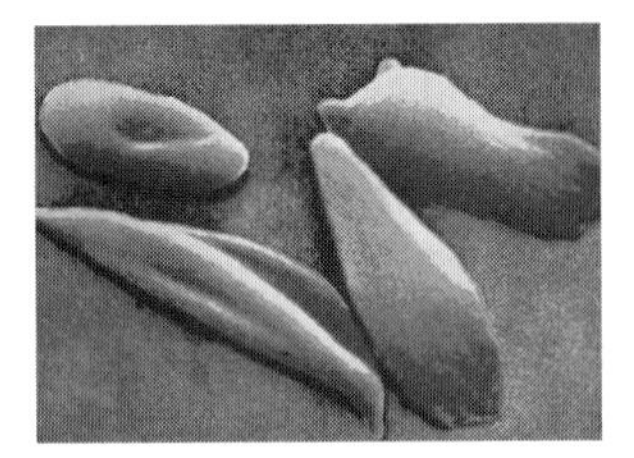

The discovery of the biochemical basis of sickle cell disease occurred at the time scientists were unravelling the secrets of the genetic code. Soon the change to the β-globulin molecule was understood at the DNA level. It was the consequence of a change to the nucleotide sequence within the globulin gene. An uncorrected nucleotide substitution had occurred within the codon representing the sixth amino acid in the globulin chain, as shown in Fig. 12.12.

CODON	*5*	*6*	*7*
normal DNA sequence	CCT	**GAG**	GAG
normal amino acid sequence	pro	**glu**	glu
sickle DNA sequence	CCT	**GTG**	GAG
sickle amino acid sequence	pro	**val**	glu

Fig. 12.12

The consequences of a nucleotide substitution in codon 6 of the β-globulin gene.

Many diseases and other conditions in living organisms are now known to be related to nucleotide substitutions. The resulting single alteration in the amino acid sequence inevitably has serious consequences for the individual, as has been illustrated by the fate of individuals with sickle cell disease. Not surprisingly, cells, and therefore individuals, rarely cope with the multiple amino acid changes that occur within a polypeptide chain, if a single nucleotide is deleted or inserted into the stored information for a particular molecule. The consequence of a deletion or of an insertion is to change all codons, and therefore all amino acids, subsequent to the mutation site. This is because the 'reading frame' of the message has shifted; thus we also refer to these classes of mutation as **frameshift mutations**. During translation the message is decoded in threes. For every adjacent three nucleotides one amino acid is added to the growing polypeptide chain. Ribosomes do not have any mechanisms to assess the validity of the message they receive. They cannot know if a message has been corrupted by the addition or removal of a nucleotide (Fig. 12.13). The effective result of a frameshift mutation is a different polypeptide, which cannot fulfil its intended function.

Ultimately we are interested in understanding the link between genotype and phenotype; how the stored information leads to the expression of different phenotypes and, indeed, of a whole organism.

Towards the end of the 1960s limitations in available technology started to become apparent. The genetic code had been elucidated and the basic features of transcription and translation worked out. Molecular geneticists wanted to examine genes in more detail, but were becoming increasingly frustrated by the lack of suitable techniques. The development in the early 1970s of a whole new methodology, known as recombinant DNA technology, retrieved the situation. Indeed it also spurred new, previously unimagined, commercial possibilities. Chapter 13 outlines the principles of recombinant technology and explores some of its applications.

normal DNA		ATCTGGCACTATGGA TAGACCGTGATACCT			
normal mRNA	AUC	UGG	CAC	UAU	CCU
polypeptide	ile	trp	his	tyr	gly
frameshift mutation			**A** inserted **T**		
mutant DNA		ATC**A**TGGCACTATGGA TAG**T**ACCGTGATACCT			
mutant mRNA	AUC	**A**UG	GCA	CUA	UGGA
polypeptide	ile	met	arg	leu	trp

Fig. 12.13

The production of a frameshift mutation.

Summary

- There is an information flow from genes to proteins.
- The information is contained within the linear nucleotide sequence of a gene. Every three adjacent nucleotides, a codon, specifies an amino acid.
- The information is carried by a mobile intermediary, messenger RNA (mRNA), to the ribosomes where proteins are synthesized.
- mRNA is one of three main types of RNA: there are also transfer RNA (tRNA) and ribosomal RNA (rRNA). All RNA molecules consist of a single polynucleotide chain.
- RNA molecules are produced by the process of transcription. One separated DNA strand acts as a template for synthesis of a complementary copy.
- rRNA is a key component of a ribosome, where assembly of amino acids into a polypeptide chain occurs. tRNA molecules bring amino acids to the ribosomes as they are required during translation.
- During translation, there is a progressive reading of the nucleotide sequence of mRNA: one amino acid being added at a time to a growing polypeptide chain.
- The nucleotide sequence of a gene is constantly surveyed by cellular enzymes for any changes to the stored information. Detected changes are corrected.
- Undetected changes represent mutations, which are of two main types – a nucleotide substitution and a frameshift.
- A mutation can alter the amino acid sequence of a protein, and thus prevent its functioning.

Problems

1. Distinguish chemically, structurally and functionally between DNA and RNA. Where is each found in the cell?

2. Name the molecules that:
   (a) display an anticodon
   (b) are synthesized by RNA polymerases
   (c) have a cloverleaf structure
   (d) possess the genetic information during protein synthesis
   (e) contain exons and introns
   (f) get charged with an amino acid

3. 5′-GGAACCCAG-3′ is the sequence of bases of a short length of DNA. Reading the sequence from left to right, give:
   (a) the base sequence that will be produced as a result of transcription of this piece of DNA
   (b) the three bases of the tRNA that will correspond to the underlined bases.

4. Why do mutations involving the deletion of a base usually have greater effects than those involving substitution of one base for another?

5. Identify the stages during transcription and translation that involve complementary base pairing.

6. Compare and contrast the structure and roles of tRNA and mRNA during translation.

7. Arrange the following terms according to their hierarchical relationship to each other: chromosomes, genomes, nucleotides, genes, codons, exons.

8. The H_1 histone protein of the toad *Xenopus laevis* is composed of 193 amino acids. What is the minimum number of nucleotides in an mRNA molecule coding for this polypeptide?

9. Match each term in the right-hand column of the table to its definition in the left-hand column.

Definition	Term
1. A group of three mRNA bases that specifies an amino acid when translated	A. Transcription
2. Most amino acids represented by several codons	B. Translation
3. Removal of introns from primary transcript	C. Codon
4. UAG, UAA or UGA	D. Colinearity
5. Insertion or deletion of a number of nucleotide pairs, other than a multiple of three	E. Reading frame
6. The process during which an RNA molecule is synthesized from a DNA template	F. Intron
7. The reading of successive mRNA base triplets as codons	G. Splicing
8. The linear correspondence between the order of amino acids in a polypeptide chain and the linear sequence of nucleotides in the encoding gene	H. Frameshift mutation
9. A nucleotide sequence that is excised from the primary mRNA transcript	I. Stop codon
10. The process during which an amino acid sequence is assembled according to the information specified by a mature mRNA	J. Degeneracy of the genetic code

10. Use Fig. 12.1 to help you complete the following table:

		G										DNA
						A	C	T				
A	G							U				mRNA transcribed
									C	U	U	tRNA anticodon
			tryptophan									encoded amino acid

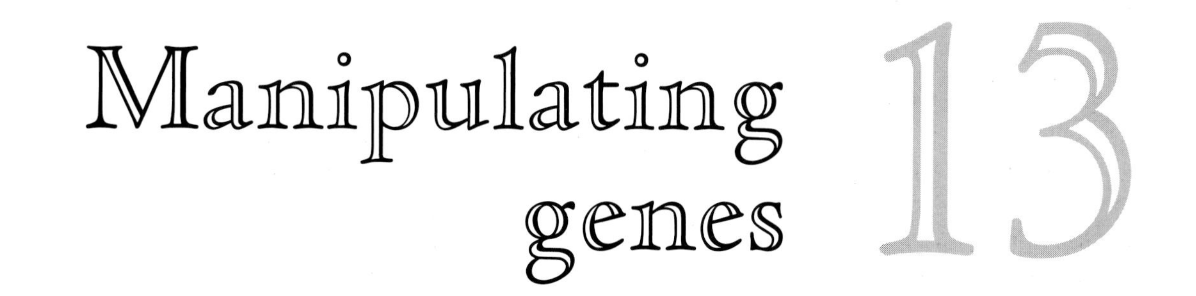

Manipulating genes 13

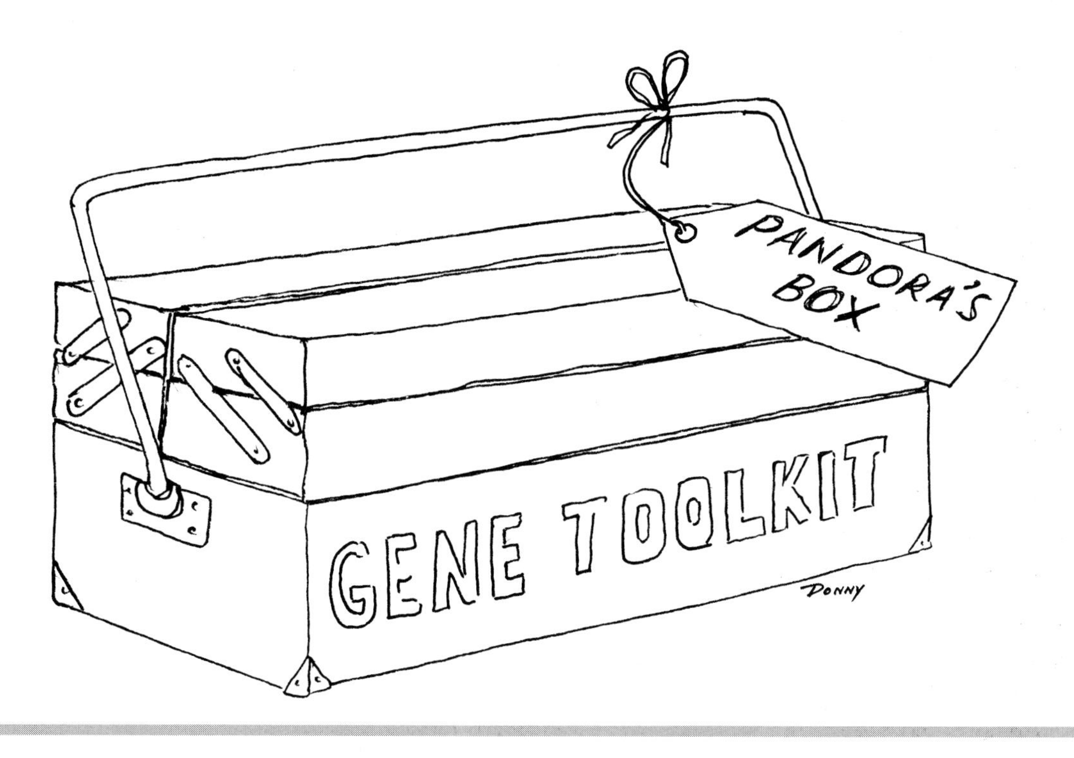

There are few things more upsetting than witnessing the pain and distress of a child suffering a haemophiliac crisis. For centuries parents have had to look on helplessly while their child bled, until eventually the blood begins to clot. Now, thanks to modern DNA technology, this need never happen. We now have the molecular 'toolkit' available to cut out a copy of the relevant gene from the genome of an individual whose blood clots properly and insert it into the DNA of non-human mammalian cells. These cells can then be cultured on a large scale and programmed to produce large quantities of the vital blood clotting protein. A haemophiliac can then have an emergency supply of the critical clotting factor constantly within easy reach and need never suffer such distress. This ability to cut and paste genes at will into the genomes of different organisms illustrates the new and rapidly expanding field of recombinant DNA technology, or 'genetic engineering'.

This chapter will introduce:

- the key methods involved in manipulating genes;
- the range of experimental and commercial uses of recombinant DNA technology.

13.1 An overview of recombinant DNA technology

Recombinant DNA technology refers to any process that modifies an organism's genotype in a directed and predetermined way. No matter if we are producing a recombinant bacterium capable of producing human growth hormone or a soybean plant resistant to the herbicide glyphosate, the approach is essentially the same and can be summarized in the following stages.

1. DNA containing the gene(s) in which we are interested is purified. This DNA is referred to as the **donor DNA**.
2. The donor DNA is chopped into approximately gene-length fragments using **restriction enzymes**.
3. The desired donor fragments are inserted into a self-replicating DNA molecule, which serves as a **vector** or carrier molecule.

   vector + inserted molecule = recombinant DNA molecule
4. The vector enters and becomes established within a host cell where either the vector is maintained as a self-replicating unit or the donor DNA enters the host cell genome.
5. As the host cell replicates, the recombinant DNA molecules are passed to progeny cells. The result is a molecular clone of the inserted fragment.
6. The final outcome depends upon the underlying purpose of the manipulation. The cloned DNA may be recovered from the host cells, purified and analysed, as for example in many genome sequencing projects. Alternatively the encoded product may be isolated and sold commercially. Harvesting clotting factor VIII for haemophiliacs or insulin for diabetics are two examples of this use of genetic engineering. If the aim of the manipulation was to confer a permanent new phenotype on the host organism, as with maize made resistant to insect attack, then the recombinant DNA acts *in situ*.

The different stages outlined above will now be considered in more detail.

13.2 Restriction enzymes

Restriction enzymes are a critical component of the cloning toolkit. These DNA 'scissors' are produced by bacteria; their role is to prevent or 'restrict' viral attack* by chopping up the infecting DNA. The discovery of restriction enzymes in the late 1960s by Werner Arber, Daniel Nathan and Hamilton Smith made molecular cloning a possibility – and earned them the 1978 Nobel prize for Physiology or Medicine.

*Bacteria, like all living organisms, are prone to attack from viruses; indeed they have their own set of infective viruses, referred to as bacteriophages.

Several hundred different restriction enzymes have now been isolated from a range of bacterial species. The enzymes are endonucleases, i.e. they cleave DNA internally. They recognize and cut at specific nucleotide sequences, generally four to six nucleotides in length. Typically a restriction site is symmetrical, i.e. the sequence is identical in both strands – though apparently reversed, because of the antiparallel nature of DNA (Fig. 13.1).

Restriction endonucleases usually make a staggered cut, generating single strands, or **sticky ends** (Fig. 13.2). Because of their complementary sequences these sticky ends have a natural affinity for each other – or similar ends. This property, combined with the fact that restriction enzymes recognize their restriction sequence whatever the source of the DNA, is the key to the use of these enzymes in recombinant DNA technology.

Fig. 13.1

Recognition sites of the two restriction enzymes Eco R1 and Hind III from the bacteria *Escherichia coli* and *Haemophilus influenzae*. The reversed symmetry of restriction sites is known as a palindrome.

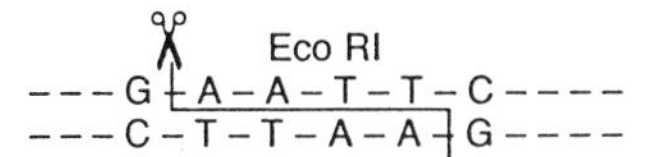

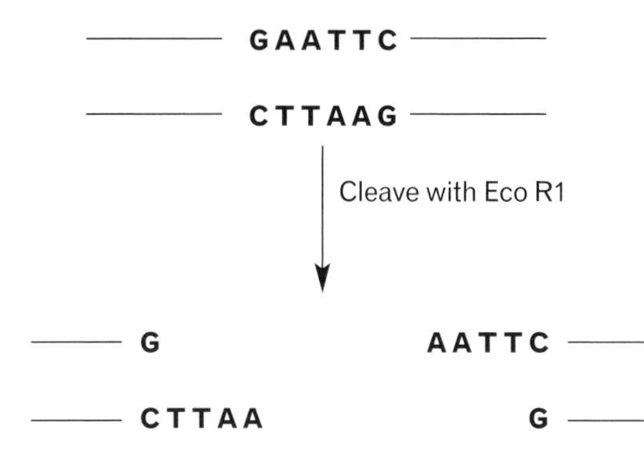

Fig. 13.2

Cleavage by the restriction enzyme Eco R1.

13.3 Vectors

Restriction enzymes enable genomic DNA to be chopped into pieces. Once the desired DNA fragment has been identified a carrier DNA molecule or **vector** is employed to introduce it into the relevant host cell. Over the last 30 years a range of vectors has been developed. The most common is a **plasmid**.

Plasmids are naturally occurring, double-stranded circular DNA molecules within the cytoplasm of bacterial and a few eukaryotic cells (Fig. 13.3). They replicate autonomously at the same time as the bacterial chromosome, producing many copies. Their precise role and origins are uncertain, but they are known to carry genes that aid the survival of bacteria under adverse conditions. For example, some plasmids have genes conferring resistance to different antibiotics. This antibiotic resistance and their autonomous replication are two characteristics that have been exploited in the development of DNA recombinant technology. The desired donor DNA fragment is inserted into a plasmid which is then introduced into the host bacterial cell. As the plasmid replicates, multiple copies of the donor DNA are produced. The antibiotic resistance represents a **marker gene**, which forms the basis of a technique signalling the presence of recombinant DNA in a bacterial cell.

The choice of cloning vector depends upon several factors; in particular, upon the size of the DNA that is being cloned and its destination. Bacterial plasmids accept relatively small fragments, just 5–10 kb in size. A human

Fig. 13.3

SEM of circular plasmid molecules isolated from *E. coli*.

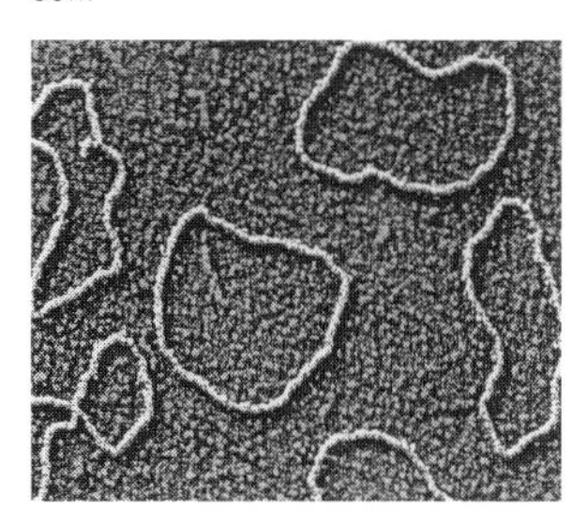

gene can be 1–2 Mb in size (1 kb = 10^3 bases and 1 Mb represents 10^6 bases). Thus, over the years, a range of different cloning vectors has been developed. Table 13.1 gives a selection, including some key eukaryotic vectors.

Table 13.1
A selection of important cloning vectors

Vector	Size of fragment	Host cells	Vector details
Plasmids	5–10 kb	Bacteria	Self-replicating double-stranded DNA molecules in bacterial cytoplasm
Lambda	15–20 kb	Bacteria	Bacteriophage
Cosmid	50 kb	Bacteria	Hybrid vector consisting of approx. one-third of the λ genome (i.e. genes controlling entry into bacterial cells) and plasmid genes for replication. Loss of most of the λ genome leaves room for bigger DNA inserts
BAC (bacterial artificial chromosome)	300 kb	Bacteria	Based on the plasmid F_1 factor
YAC (yeast artificial chromosome)	1–2 Mb	Yeast and mammalian cells	Hybrid chromosome containing a centromere, telomeres at either end, an origin of replication, a cluster of restriction sites for insertion of donor DNA and selectable marker genes
Ti plasmid	180 kb	Crop plants	Plasmid of soil bacterium, *Agrobacterium tumefaciens*. A segment, T-DNA, readily inserts into host chromosomes, carrying donor genes
Retrovirus	8 kb	Mammalian cells	Its single-stranded RNA, which also contains a transcript of donor DNA, is reverse transcribed into DNA within a host cell and integrates into host chromosomes

Bacteriophages are the viruses of bacteria. Their value in cloning is that they have the genetic apparatus to infect and multiply within a bacteria cell. Commonly lambda (λ) phage (Fig. 13.4) or a hybrid version, a cosmid, is used as a cloning vector. Cosmids and BACs have proved useful cloning vectors for the human genome project. DNA recombinant techniques were developed using prokaryotic systems. However, if we are interested in the expression of eukaryotic genes, a eukaryotic cloning organism is more desirable. Yeast, *Saccharomyces cerevisiae*, is an ideal eukaryotic host species (Fig. 13.5). It is unicellular, and so can be grown and manipulated in a similar way to bacterial cells. Much is known of its genetic system and it contains a plasmid! A linear yeast artificial chromosome, or **YAC**, has also been created and widely used as a cloning vector.

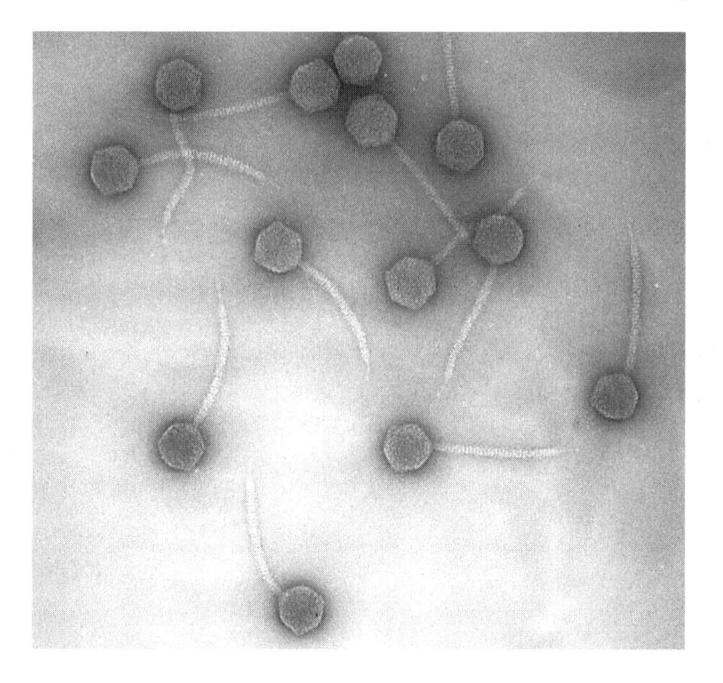

Fig. 13.4
Electron micrograph of Lambda bacteriophages.

13.4 Creating a recombinant DNA molecule

Restriction enzymes are again used in the production of the recombinant plasmid. A cloning plasmid is chosen that contains the nucleotide sequence recognized by the restriction enzyme that was used to produce the donor DNA fragments. Alternatively relevant recognition sequences can now be engineered into a plasmid. The chosen plasmid is treated with the restriction enzyme. This opens up, or **linearizes**, the plasmid and produces sticky ends. These sticky ends will be the same as those of the donor DNA. When the donor fragments and linearized plasmid are mixed, complementary base pairing occurs between the sticky-ended single strands produced by the restriction enzyme. DNA ligase is added to seal the joints. The process is represented in Fig. 13.6.

Fig. 13.5
Scanning electron micrograph of *Saccharomyces cerevisiae*. Cells can be seen reproducing asexually by growth of buds.

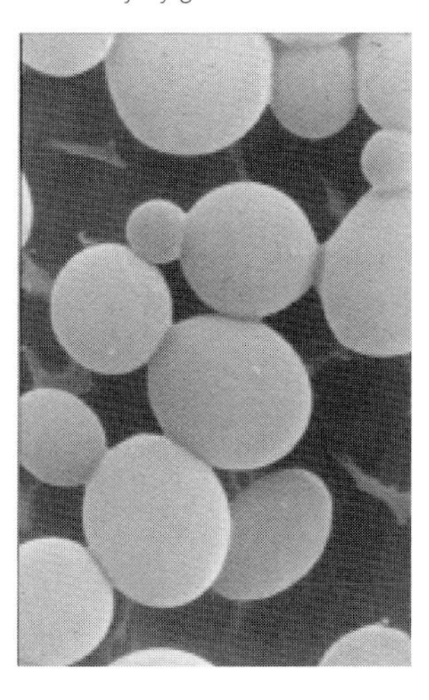

13.5 Introducing the recombinant DNA molecule into a host cell

Once the recombinant DNA molecules have been produced they are introduced into host cells. Various techniques exist. Simplest is the treatment of bacteria cells with calcium chloride to facilitate the uptake of plasmids. Alternatively, a process called **electroporation** can be used, when cells are stimulated with a brief, weak electric shock. This causes the cell membrane to become temporarily permeable to DNA. Bacterial cells that have taken up

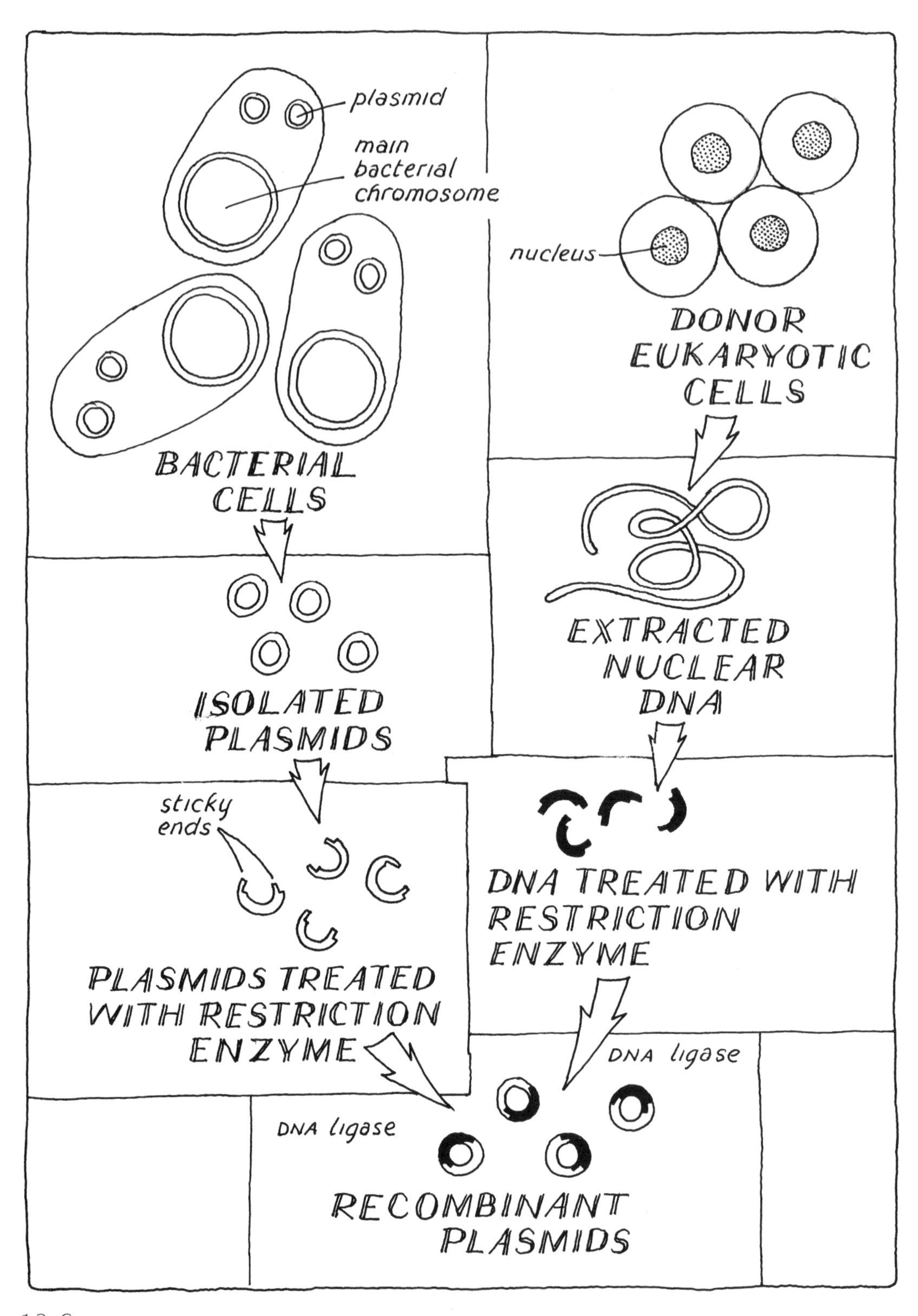

Fig. 13.6
Construction of recombinant plasmids.

recombinant DNA are referred to as **transformed cells**. Transformed cells have to be identified and separated from those cells that lack a plasmid. A common way to distinguish between the two types of cells makes use of the antibiotic resistance conferred on a host cell by the presence of a plasmid.

Bacteria can be grown in liquid culture or on solid nutrient plates where they form **colonies**. All bacterial cells within a colony are derived from a single ancestral cell: thus all cells of a colony, and the plasmid they contain, are genetically identical. To screen for those cells that have taken up a plasmid, the bacteria can be plated on antibiotic-containing agar. Only those bacteria with a plasmid, and so a resistance gene, will grow.

The use made of the transformed cells with their recombinant DNA varies according to the host organism and, of course, the reasons for their production. In 1973 Herbert Boyer and Stanley Cohen performed the first successful recombinant DNA cloning experiment. Thirty years later the range of uses of recombinant technology is enormous. It has revolutionized all fields of experimental biology and spread far beyond the research laboratory into the commercial market, supporting a booming biotechnology industry. The following list gives a selection of the range of uses of recombinant DNA technology:

- The efficient production of useful proteins in large quantities; for example, clotting factor VIII, human growth hormone and insulin, with their obvious medical applications, or rennin for the cheese industry.
- Generation of DNA sequences as research tools or for use in medical diagnosis such as prenatal diagnosis or carrier testing.
- The introduction of genes into plants and animals to produce new, desired characteristics; for example human genes into the genome of pigs as part of the programme to produce hearts for transplantation into humans, or herbicide resistance genes into the genomes of many crop plants.
- The correction of genetic defects by introducing fully functional genes into cells, i.e. gene therapy.

Although the details may differ, the overall approach to achieving any of the outcomes in the above list is remarkably uniform. The use of recombinant techniques in medicine and agriculture perhaps generates the fiercest debate; these two areas will be discussed further.

13.6 The production of human recombinant proteins

Initially biotechnology companies employed bacteria to produce their recombinant proteins. The first human gene product manufactured in this way and licensed for therapeutic use was insulin, which became available in 1982. Many other useful recombinant proteins have followed (Table 13.2). In the early 1980s it seemed almost miraculous that the means had been found of providing, relatively cheaply, large supplies of much needed medicines. Soon, however, disturbing reports began to emerge. Some individuals using engineered insulin and growth hormone were suffering unpleasant side-effects. It seemed that there were imperfections in the system. Eukaryotic proteins did not always fold properly into their correct three-dimensional form in the prokaryotic bacterial cytoplasm. Bacterial cells could not fully process and modify the recombinant protein if extra groups, such as phosphates or sugars, needed to be added post-translationally. Such discoveries suggested a new approach was needed to the production of recombinant proteins. It was argued that eukaryotic recombinant proteins should be produced by eukaryotic cells; and, if possible, human proteins by mammalian cells. Attention was focused on the possibility of modifying the genome of livestock so that they produced much needed human proteins in their milk. One of the first proteins to be successfully produced in this way was the enzyme alpha-1-antitrypsin, in the milk of Tracy the sheep. Lack of alpha-antitrypsin produces a hereditary form of emphysema that is progressive and eventually fatal.

Table 13.2

A selection of human recombinant protein available or in clinical testing

Gene product	Condition treated
Epidermal growth factor	Burns and skin grafts
Erythropoietin	Anaemia
Factor VIII	Haemophilia A
Gamma interferon	Cancer
Hepatitis B vaccine	Hepatitis B
Human growth hormone	Dwarfism
Insulin	Diabetes
Interleukin-2	Cancer
Tissue plasminogen activator	Heart attacks

Two major problems were faced by geneticists when developing these new recombinant technologies:

- how to introduce the relevant human gene into the genome of a sheep or other mammal;
- how to limit expression of the introduced gene to the mammary tissue.

It was soon realized that the best time to introduce foreign genes into the genome of a complex multicellular organism such as a sheep, was at a very early embryological stage, even into the zygote or unfertilized egg. In the case of Tracy the alpha-1-antitrypsin gene was introduced into the zygote. To ensure that expression of the gene only occurred in mammary tissue, the antitrypsin gene was hitched to a special control region that is only operational in this tissue. The original bacterial recombinant technology still had its uses. Bacteria were used to provide multiple copies of the fusion gene. Figure 13.7 summarizes the procedure used to produce Tracy. When the transformed zygote had divided a few times the embryo was introduced into a foster mother. Once sexually mature Tracy was mated with a normal ram. The milk she produced had high concentrations of the protein.

Cattle, sheep and goats are now routinely used as animal biofactories. Another sheep, Polly, was the first to produce human clotting factor IX. The cloning of the now famous Dolly was viewed as the first stage in a long-term programme aimed at developing techniques to clone efficient human protein producers such as Tracy and Polly. The rationale was that, once you had a successful protein producing animal it is easier to make multiple copies of this animal than to start the transformation process again! However, the creation of Dolly initiated strong debate because the technology used in her creation could, theoretically, be used in attempts to clone humans.

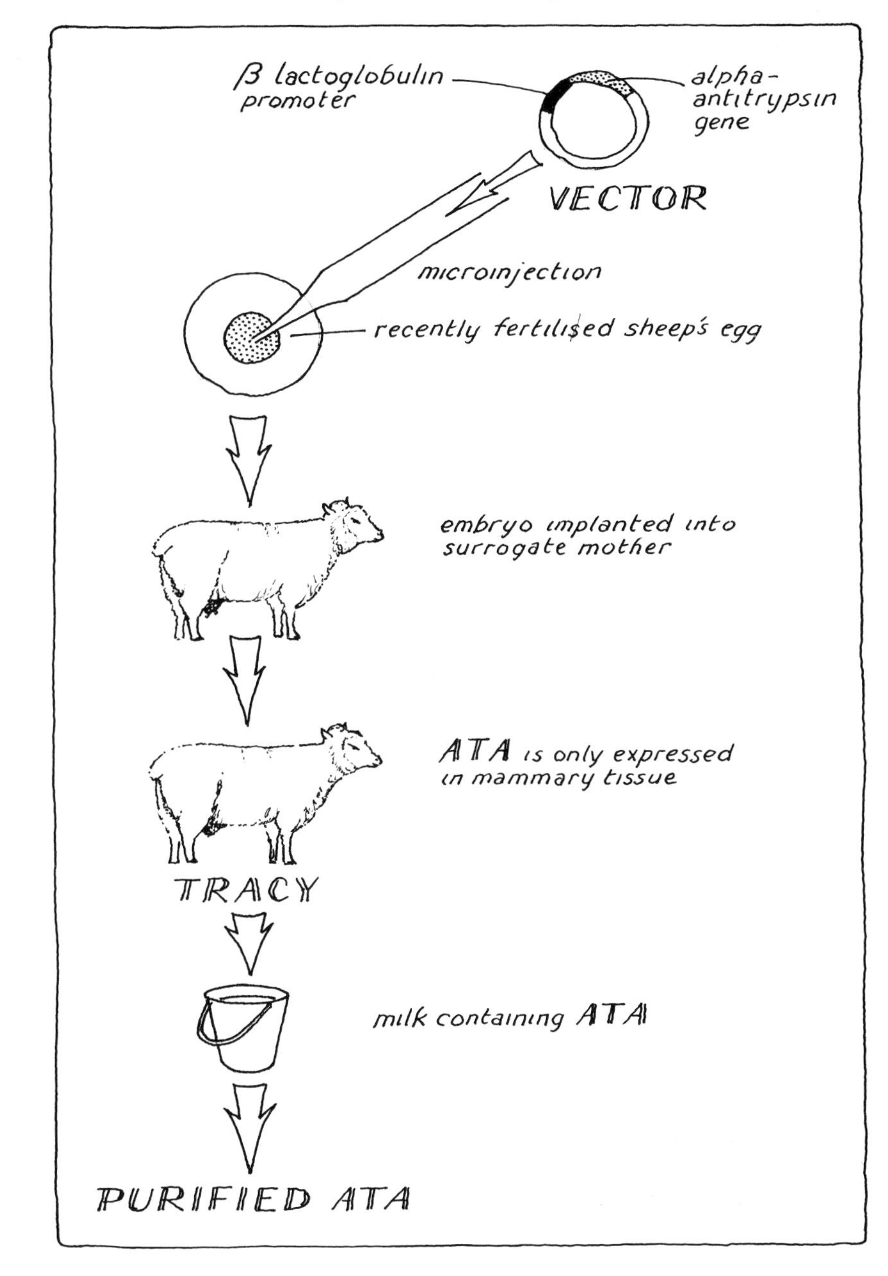

Fig. 13.7 The creation of Tracy. The alpha-antitrypsin (ATA) gene was linked to the β-lactoglobulin promoter. Because this promoter is only expressed in mammary glands, ATA production is limited to this tissue. The protein is secreted into the milk, from which it is purified.

Production of a genetically modified egg, zygote or early embryo forms the basis of producing **transgenic animals** such as Tracy and Polly. Transgenic animals are not, however, only produced with the purpose of harvesting high yields of a recombinant protein. Over the last decade scientists have devoted much effort towards producing transgenic pigs as a potential source of hearts and other organs for transplantation into humans. The transgenic pigs express human proteins within the membranes of their cells. The hope is that such transplanted tissue will not be recognized by the ever alert human immune system, which initiates rejection of foreign tissue. Transgenic mice have become model systems for the study of a range of human diseases. We are even, in the early years of the twenty-first century, allowing ourselves to contemplate the possibility of producing 'transgenic humans' as a means of curing genetic disease.

Hopes were high in the late 1980s that many serious inherited disorders could be cured by replacing deficient genes with functional copies. The results, however, of a decade of intensive research and clinical trials have been disappointing. It has proved enormously difficult to successfully deliver the modifying genes into the correct target tissue, let alone ensure the genes' integration into a suitable human chromosome where its long-term expression could occur. Could germ line therapy (i.e. altering all cells including those that give rise to the gametes) be the solution to this seemingly intractable problem? The implications of modifying the genome of a germ line cell are enormous. The cells are irrevocably changed. We might achieve our goal of curing fatal genetic disease but with it we are likely to raise all sorts of other, unforeseen developments. How might we, for example, respond to the suggestion that we use germ line gene therapy to extend intellectual potential or enhance athletic ability? The difficult ethical dilemmas created by the development of recombinant techniques are not confined to transgenic animals. Some of the most intense debate of recent years regarding the production of genetically modified organisms has focused on the development of transgenic crops.

13.7 The production of transgenic crops

Weed infestation is responsible for an estimated 10% decrease in crop productivity world-wide. The response to this problem in the recent past has been to use a vast range of different herbicides to kill these offending weeds. But there has always been the risk of killing the crop itself. Furthermore, many herbicides are long-lasting. They seep into water supplies, sometimes creating toxicity problems for humans. One of the few herbicides that is effective at low concentrations, short-lived (because it is degraded by soil microorganisms) and non-toxic to humans is glyphosate. Unfortunately, crop susceptibility was too high, until recombinant DNA technology came up with a solution.

Glyphosate kills plants because it inhibits the action of a crucial chloroplast enzyme, EPSP synthetase, which is important in amino acid biosynthesis.

Certain strains of *E. coli* have a changed EPSP synthetase enzyme and are resistant to the inhibitory effects of glyphosate. Why not put a copy of this bacterial gene into the genome of different crop plants? This is just what Monsanto and other biotech companies achieved in the early 1990s. An ideal system, using the plasmid present in the soil bacterium *Agrobacterium tumefaciens*, was already available for transferring genes into plant genomes.

Agrobacterium tumefaciens naturally enters plants at sites of wounds, stimulating the formation of calluses. *A. tumefaciens,* like many bacteria, possesses a plasmid. It is this $\mathbf{T_1}$ plasmid that is responsible for the development of a callous, or to be more precise, one region of the plasmid genome, the **T DNA**. When plants are infected with *A. tumefaciens* the T DNA inserts into one of the plant's chromosomes and directs the synthesis of cell division proteins. The recombinant DNA strategy is to hitch the donor gene(s) to the T DNA, which integrates them into the plant genome.

Cells transformed with the T_1 plasmid need to be identified. For this reason the original modification usually occurs in culture. For example *A. tumefaciens* containing the EPSP resistant gene within its T_1 plasmid infects plant discs (Fig. 13.8). When a callous appears the cells are then cultured on medium containing glyphosate. Any cells that grow must have the resistant gene. Techniques are available to nurture these undifferentiated cells to produce root and stem systems. In addition to the production of transgenic maize, soybean and wheat resistant to the herbicide glyphosate, various other commercially useful genes have been introduced into a range of different crop plants (Table 13.3).

Table 13.3

A selection of the range of different traits that have been introduced into plants by genetic manipulation

Trait	Modified plant
Insect resistance	Maize, cotton, potato, tomato
Herbicide resistance	Maize, soybean, cotton, rape, sugar beet, rice, flax
Virus resistance	Squash, papaya, potato
Delayed fruit ripening	Tomato
Altered oil content	Rape, soybean
Hepatitis B vaccine	Tobacco
Pollen control	Maize, chicory

The scientific success in producing these genetically modified crops has not been matched by commercial success. Concerns soon started to be expressed over the production and uses of these crops: over the safety of foods derived from genetically modified (GM) crops and their environmental impact. As a result no GM crops are currently grown on a commercial scale in the UK. In 2000 just 13 countries world-wide were growing GM crops, with 99% of this production concentrated in Argentina, Canada, China and the USA. Table 13.4 contrasts some of the key concerns over the growth of GM crops with what might be regarded as the benefits.

The verdict remains open on the safety and propriety of the development and use of GM crops. There seems little doubt that scientists, farmers, politicians, retailers, environmental groups, consumers and others will continue debating the issues for many years yet. Indeed, as we develop an ever more detailed understanding of how genes work and how expression is coordinated on a cell-to-cell, day-to-day and life-span basis, we will invent new, hitherto undreamt-of applications with an attendant spectrum of difficult ethical and social issues. Can all our knowledge also bring us closer to understanding what it means to be human? Our response will, perhaps, depend upon our personal, philosophical or religious beliefs. Who could have predicted that Gregor Mendel's patient observations in the monastery garden at Brno could have led, in a little more than a century, to the rich and diverse science of today?

Summary

- Genes can be manipulated between a wide range of species – bacteria, fungi, animals and plants.
- Restriction enzymes cut the donor DNA at specific sites. The relevant fragment is inserted into a vector that introduces the foreign DNA into cells of the host species.
- Recombinant DNA technology has a wide spectrum of uses. These include producing useful proteins in large quantities, producing transgenic animals and plants and human gene therapy.
- All forms of gene manipulation raise issues of safety and ethical acceptability.

Glossary

Allele: one of two or more forms of a gene.
Alloploid: an individual, or cell, which possesses two or more distinct chromosome sets derived from different species.
Amino acids: the covalently linked building blocks of proteins.
Anaphase: the stage of nuclear division during which homologous chromosomes or sister chromatids separate and move towards opposite poles of the cell.
Aneuploid: an individual, or cell, in which the chromosome number is not an exact multiple of the haploid set.
Anticodon: the nucleotide triplet in a tRNA molecule that is complementary to an mRNA codon.
Autopolyploid: an individual, or cell, which possesses three or more identical sets of chromosomes.
Autosome: any chromosome other than a sex-determining chromosome.
Backcross: a cross that involves an F_1 heterozygote and one of the parents; generally the recessive one.
Bacteriophage: a virus that infects bacteria.
Barr body: an inactive X chromosome, visible as a densely staining mass within the somatic nucleus of mammalian females.
Bivalent: a pair of associated homologous chromosomes during prophase I of meiosis.
Bottleneck: a transitory shrinking of population size when alleles may be lost from the gene pool.
Carrier: an individual who is heterozygous for a recessive trait.
Cell cycle: the sequential phases of growth of an individual cell; starting with G1 (gap 1) followed by S (DNA synthesis), G2 (gap 2) and, finally, M (mitosis).
Centimorgan (cM): a unit of distance between genes on a chromosome; one cM corresponds to a recombination frequency of 1%.
Centrioles: a structure, consisting of a collection of tiny microtubules, around which the spindle is organized during mitosis and meiosis.
Centromere: specialized region of a chromosome to which the spindle fibres attach during nuclear division.
Chiasma: the location of a **crossover** between non-sister chromatids during prophase I of meiosis.
Chromatid: one of the two longitudinal subunits of a replicated chromosome, joined to its sister chromatid at the **centromere**.
Chromatin: the complex of **histone** proteins and DNA that make up chromosomes.

Chromosome: a thread-like structure consisting of **chromatin**, containing genetic information arranged in a linear sequence.

Codominance: condition in which the phenotypic effects of a pair of alleles are fully and simultaneously expressed in a heterozygote.

Codon: a triplet of nucleotides in a DNA or RNA molecule that specifies an amino acid.

Complementary base pairing: a chemical affinity between nitrogenous bases, such that adenine pairs only with thymine and guanine only with cytosine.

Crossing over: the exchange of genetic material between homologous chromosomes during prophase I of meiosis; it produces **recombination**.

Cytokinesis: cell division that follows mitosis and meiosis.

Deletion: the loss of chromosomal material, ranging from a single nucleotide to many genes.

Dihybrid cross: a genetic cross in which the parents possess different forms of two traits.

Diploid: an individual, or cell, which possesses two copies of each chromosome.

DNA (deoxyribonucleic acid): a macromolecule consisting of two antiparallel polynucleotide chains held together by hydrogen bonds; the primary carrier of genetic information.

DNA ligase: an enzyme that catalyses the formation of a phosphodiester bond.

DNA polymerase: an enzyme that catalyses the synthesis of DNA from deoxynucleotides and a DNA template.

Dominance: a condition in which the phenotypic effect of only one of a pair of alleles is expressed in a heterozygote.

Double crossover: two separate exchanges of genetic material between homologous chromosomes during prophase I of meiosis.

Double helix: describes the helical configuration of the two antiparallel polynucleotide chains of DNA.

Duplication: the presence of a repeated segment of a chromosome.

Enzyme: a protein that catalyses a specific biochemical reaction.

Epistasis: a non-reciprocal interaction between genes such that an allele at one gene interferes with the expression of alleles at another gene.

Euchromatin: lightly staining chromosomal regions that are transcriptionally active during interphase.

Eukaryote: an organism whose cells possess a true nucleus and membranous organelles.

Euploid: an individual, or cell, with one or more complete sets of chromosomes.

Exons: the coding portions of genes that are transcribed and translated into protein.

F_1 generation: the first filial generation, or progeny resulting from the first cross in a series.

F_2 generation: the second filial generation, or progeny resulting from a cross involving the F_1 generation.

Fixation: a condition in which all members of a population are homozygous for a given allele.

Founder effect: the establishment of a population by a small group of individuals whose genotypes carry only a fraction of the different alleles in the parental population: genetic drift acts to alter allele frequencies.

Frameshift mutation: the insertion or removal of one or more nucleotides into or from a gene, which shifts the codon reading frame in all codons following the change.

Gamete: the haploid reproductive cell (sperm or ovum).

Gene: the fundamental unit of heredity that occupies a specific chromosomal **locus**.

Gene flow: the gradual exchange of genes between two populations brought about by the dispersal of gametes or the migration of individuals.

Gene pool: a common set of genes, and their alleles, shared by a group of interbreeding individuals.

Genetic drift: the random variation in allele frequency from generation to generation; most often observed in small populations.

Genetic equilibrium: the maintenance of allele frequencies at a constant value in successive generations, if no outside forces are acting on them.

Genome: the totality of an organism's DNA.

Genotype: the full complement of an individual's genes **or** the allelic composition of one or a few genes.

Germ line: gamete-producing cells.

Giemsa stain: a type of stain that produces G bands in chromosomes.

Haploid: an individual, or cell, that possesses one copy of each chromosome.

Hardy–Weinberg law: the principle that allele and genotype frequencies will remain constant in succeeding generations in an infinitely large population and in the absence of mutation, migration, selection and non-random mating.

Hemizygous: a condition in which only a single copy of a gene is present; usually applies to genes on the X chromosomes in **heterogametic** males.

Heritability: the proportion of the variability of a trait attributable to genetic factors.

Heterochromatin: heavily staining chromosomal regions that are condensed and transcriptionally inactive.

Heterogametic sex: sex that produces gametes containing different sex chromosomes.

Heterozygote: an individual with two different alleles at a locus.

Histone: the protein around which DNA is wound in a chromosome.

Homogametic sex: sex that produces gametes containing identical sex chromosomes.

Homologues: chromosomes that possess identical loci, and pair during meiosis.

Homozygote: an individual with two identical alleles at a locus.

Inbreeding: mating between closely related organisms.

Incomplete dominance: expression of a heterozygous phenotype that is distinct from, and often intermediate to, the phenotype produced by either homozygote.

Incomplete penetrance: the possession of a mutant phenotype without expressing the corresponding mutant phenotype.

Interphase: the portion of the cell cycle between cell divisions.
Intron: the non-coding portion of a gene, which is transcribed and subsequently removed from the mRNA molecule.
Inversion: the reversal of a segment of DNA.
Karyotype: a display of a cell's chromosomes organized according to size.
Lethal allele: an allele whose expression results in death of an individual.
Linkage: two loci that are situated close together on the same chromosome.
Locus: the chromosomal location of a specific gene.
Mean: the arithmetic average.
Meiosis: a process of nuclear division during which haploid cells are formed from diploid ones.
Messenger RNA (mRNA): RNA molecules that are transcribed from DNA and translated into an amino acid sequence.
Metaphase: the stage of nuclear division during which condensed chromosomes lie in the central plane between the two poles of a cell.
Mitosis: a process of nuclear division that produces cells with the same chromosome number and genetic complement as the parental one.
Monohybrid cross: a genetic cross in which the parents possess different forms of one trait.
Monomorphism: the existence of only one allele at a locus.
Monosomy: an aneuploid condition in which one member of a pair of chromosomes is absent.
Multifactorial trait: a trait that is the result of interactions of multiple genetic and environmental factors.
Multiple alleles: three or more alleles of one gene.
Mutant: an individual, or cell, carrying an altered gene.
Mutation: an alteration in gene or chromosome structure.
Non-disjunction: failure of homologous chromosomes or sister chromatids to segregate into separate cells during meiosis or mitosis.
Nucleosome: a structural unit of eukaryotic chromosomes, formed by 150 bp of DNA wrapped around a histone core.
Nucleotides: the covalently linked building blocks of DNA and RNA.
Okazaki fragment: small segment of DNA produced during DNA replication.
Pedigree: a diagram that shows family relationships and transmission of genetic traits over several generations.
Phenotype: the observed characteristics of an individual produced by specific genotypes.
Phosphodiester bond: the covalent link between adjacent nucleotides in DNA and RNA.
Plasmid: circular, double-stranded DNA molecule within the cytoplasm of prokaryotic and a few eukaryotic cells. Capable of independent replication and often used as a cloning vector in recombinant DNA techniques.
Pleiotropy: describes genes that have multiple phenotypic effects.
Polygenic: describes a trait determined by the additive effect of many genes.
Polymer: a macromolecule produced by multiple copies of a simple molecule.
Polymorphism: the existence of two or more different alleles at a locus, the frequency of the most common being 0.99 or less.

Polyploid: an individual, or cell, that possesses three or more sets of chromosomes.

Polysome: a structure consisting of two or more ribosomes engaged in the translation of a similar number of mRNA molecules.

Population: a local group of a single species within which mating is actually or potentially occurring.

Prokaryote: an organism whose cells lack a true nucleus and membranous organelles.

Prophase: the first stage of nuclear division.

Protein: a molecule composed of one or more polypeptides, each consisting of covalently linked amino acids.

Punnett Square: a table which specifies all the possible genotypes that can result from fertilizations between the gametes of a pair of mating individuals.

Pure-breeding: see true breeding.

Quantitative trait: a characteristic that can be measured on a continuous scale and whose expression is the result of the additive effect of many genes and environmental factors.

Recessive: an allele that is only phenotypically expressed in the homozygous state.

Reciprocal cross: a paired cross in which the genotype of the female in the first cross is represented as the genotype of the male in the second cross, and vice versa.

Recombinant DNA: a DNA molecule that consists of DNA from more than one parent molecule, e.g. human DNA inserted into a cloning vector.

Recombination: the process that leads to new combinations of alleles, as the result of crossovers during parental meiosis.

Restriction enzyme: a bacterial enzyme that cleaves double-stranded DNA at a specific nucleotide sequence (restriction site).

Ribosomal RNA (rRNA): RNA molecules that, along with protein molecules, are structural components of ribosomes.

Ribosome: the cytoplasmic site of translation of mRNA into amino acid sequences.

RNA polymerase: an enzyme that catalyses the synthesis of RNA from a DNA template.

Selection: the force that brings about changes in the frequency of alleles and genotypes within a population through differential reproduction.

Selection coefficient(s): a measurement of the reproductive disadvantage of a given genotype in a population. The higher its value, the more disadvantageous the genotype.

Selfing: reproduction by self-fertilization.

Semi-conservative replication: a model of DNA replication in which each daughter molecule consists of an original parental strand and a newly synthesized one.

Sex chromosome: a chromosome, such as the X and Y mammalian chromosomes, that is involved in determining the sex of an individual.

Sex linkage: refers to genes located on a sex chromosome.

Solenoid: a structural unit of eukaryotic chromosomes consisting of approximately six **nucleosomes**.

Somatic cells: all cells other than **germ cells**.

Standard deviation: a quantitative measure of the amount of variation shown by a population: it indicates the percentage of measurements within a certain range of the **mean**.

Sticky end: the single-stranded extension of a double-stranded DNA molecule.

Telophase: the stage of nuclear division during which separated chromosomes or chromatids reach opposite poles, and new nuclear envelopes form.

Test cross: a cross between an individual expressing the dominant phenotype, but of unknown genotype, and a homozygous recessive individual.

Tetraploid: an individual, or cell, that possesses four sets of chromosomes.

Transcription: the process during which an RNA molecule is synthesized from a DNA template.

Translation: the process during which an amino acid sequence is assembled, at a ribosome, according to the specific nucleotide sequence of an mRNA molecule.

Translocation: the exchange of genetic material between two non-homologous chromosomes.

Triploid: an individual, or cell, that possesses three sets of chromosomes.

Trisomy: an aneuploid condition in which an extra copy of one chromosome is present.

Transfer RNA (tRNA): an RNA molecule that brings specific amino acids to the ribosome, as required during **translation**.

Transformation: the acquisition by a cell of new genes.

Transgenic organism: an organism that has had its genotype altered by the introduction of new gene(s) into its genome by genetic manipulation.

True breeding: a variety or strain that yields progeny like itself (homozygous).

Variance: a quantitative measure of the variation of values from a mean.

Vector: the vehicle (e.g. plasmid, cosmid, BAC or YAC) used to carry a DNA insert into a host cell.

Further reading

There is a vast and wide-ranging genetics literature to be found, both on the World Wide Web and in print. This list includes a selection of general genetics texts that develop further many of the ideas introduced in this book, a number of books that focus on specific topics and a couple about Gregor Mendel and the origins of genetics.

Brooker, R.J. (1999) *Genetic Analysis & Principles*. Addison, Wesley & Longman, London.

Brown, T.A. (1998) *Genetics: a Molecular Approach*, 3rd edn. Nelson Thornes, Cheltenham.

Connor, M. and Ferguson-Smith, M. (1997) *Essential Medical Genetics*, 5th edn. Blackwell, Oxford.

Falconer, D.S., *et al.* (1996) *Introduction to Quantitative Genetics*, 3rd edn. Longman, London.

Frankham, R., *et al.* (2002) *An Introduction to Conservation Genetics*. CUP, Cambridge.

Griffiths, A.J.F., *et al.* (2002) *Modern Genetic Analysis*, 2nd edn. Freeman, New York.

Hartl, D. (2000) *A Primer of Population Genetics*, 3rd edn. Sinauer Associates, Sunderland, USA.

Hartl, D.L. and Jones, E.W. (2000) *Genetics: an Analysis of Genes and Genomes*, 5th edn. Jones & Bartlett Pubs, Sudbury, USA.

Hartwell, L.H., *et al.* (2000) *Genetics: from Genes to Genomes*. McGraw-Hill, New York.

Jorde, L.B., *et al.* (2000) *Medical Genetics*. Mosby, St Louis, USA.

King, R. and Stansfield, W. (1996) *A Dictionary of Genetics*, 5th edn. OUP, Oxford.

Klug, W. and Cummings, M.R. (2000) *Concepts of Genetics*, 6th edn. Prentice Hall, Upper Saddle River, USA.

Lewin, B. (2000) *Genes VII*, 7th edn. OUP, Oxford.

Mange, E.J. and Mange, A.P. (1998) *Basic Human Genetics*, Sinauer Associates, Sunderland, USA.

Marantz Henig, R. (2000) *The Monk in the Garden: The Lost and Found Genius of Gregor Mendel.* Houghton Mifflin, Boston.

Olby, R. (1985) *Origins of Mendelism*, 2nd edn. University of Chicago Press, London.

Orel, V. (1995) *Gregor Mendel: The First Geneticist.* OUP, Oxford.

Plomin, R., *et al.* (2001) *Behavioural Genetics*, 4th edn. Worth Publishers, New York.

Reiss, M.J. and Straughan, R. (2001) *Improving Nature?: The Science and Ethics of Genetic Engineering.* CUP, Cambridge.

Russell, P. (2002) *Genetics*. Benjamin Cummings, San Francisco.

Tamarin, R.H. (2001) *Principles of Genetics*, 7th edn. McGraw-Hill, Boston.

Solutions to problems

Chapter 2

1. Let **B** = allele for black eyes, **b** = allele for red eyes
   (a) F_1 all black-eyed
   black-eyed male (**BB**) × red-eyed female (**bb**) → all black-eyed (**Bb**)
   (b) Three black-eyed mice to every one red-eyed mouse.
   F_1 black-eyed (**Bb**) × F_1 black-eyed (**Bb**) → 1 **BB** (black-eyed) : 2 **Bb** (black-eyed) : 1 **bb** (red-eyed)
2. Perform a test cross, i.e. cross the black-eyed mouse with a red-eyed one. The presence of red-eyed mice among the test cross progeny indicates the F_2 black-eyed mouse was heterozygous.
   If black-eyed mouse was **BB**: **BB** × **bb** → **Bb** (black-eyed)
   If black-eyed mouse was **Bb**: **Bb** × **bb** → 50% **Bb** (black-eyed); 50% **bb** (red-eyed)
3. (a) The results suggest that hair length is determined by a single gene with two alleles – one promoting the growth of short hair and the other long hair. The 3 : 1 offspring ratio of short- to long-haired guinea pigs indicates that short hair is the dominant trait.
   (b) Heterozygous, **Ss** (if **S** = allele for short hair and **s** = allele for long hair). A 3 : 1 offspring ratio indicates heterozygous parents.
4. The results suggest that polydactyly is determined by a dominant allele at a single locus.
5. The F_1 poppies were spotted. A 3 : 1 offspring ratio indicates spotted is the dominant trait and that F_1 parents are heterozygous.
6. All their children will have dimples:
   dimpled father (**DD**) × mother lacking dimples (**dd**) → all children (**Dd**)
7. The 3 : 1 ratio among the progeny produced by crossing flies Q and R indicates that one gene with two alleles determines body colour with the grey body allele dominant to the black body allele. Flies Q and R are therefore heterozygous, while fly P is homozygous dominant. Thus, if **G** = allele for grey body and **g** = allele for black body, the results of crossing the three flies with a black-bodied one are:
   fly P (**GG**) × black-bodied fly (**gg**) → 100% grey-bodied flies (**Gg**)
   fly Q or fly R (**Gg**) × black-bodied fly (**gg**) → 50% grey-bodied flies (**Gg**); 50% black-bodied flies (**gg**)
8. 480 plants should be large-leaved and 160 plants to be small-leaved. The F_1 results indicate that large leaves is the dominant trait. The F_1 plants are heterozygous. Crossing two heterozygous individuals produces a progeny ratio of 3 : 1 dominant to recessive individuals.
9. Nil.
10. Shell thickness is controlled by a single gene with two alleles; one (**T**) promotes the development of thick shells and the other (**t**) promotes the development of thin shells.
    tree A (**Tt**) × thin-shelled tree (**tt**) → 50% **Tt** (thick-shelled); 50% **tt** (thin-shelled)
    tree B (**TT**) × thin-shelled tree (**tt**) → 100% **Tt** (thick-shelled)
11. (a) Heterozygous, **Pp**.
    polled (**Pp**) × polled (**Pp**) → 1 **PP** (polled) : 2 **Pp** (polled) : 1 **pp** (horned)
    (b) (i) ¾ or 75% (genotypes **PP** and **Pp**); (ii) ¼ or 25% (genotype **pp**); (iii) ¼ or 25% (genotype **PP**)

Chapter 3

1. Let **R** = allele for red flowers and **W** = allele for white flowers
   Genotypes: red-flowering plants = **RR**; pink-flowering plants = **RW**; white-flowering plants = **WW**
   (a) 25% red : 50% pink : 25% white-flowering plants
   pink (**RW**) × pink (**RW**) → 1 **RR** (red) : 2**RW** (pink) : 1 **WW** (white)
   (b) 50% white : 50% pink-flowering plants
   WW (white) × **RW** (pink) → 1 **WW** (white) : 1 **RW** (pink)

(c) all pink-flowering plants

WW (white) × **RR** (red) → **RW** (pink)

(d) all white-flowering plants

WW (white) × **WW** (white) → **WW** (white)

2. Henrik's parents have the genotypes $\mathbf{I^AI^i}$ and $\mathbf{I^BI^i}$. Because Henrik is blood group B his genotype must be $\mathbf{I^BI^B}$ or $\mathbf{I^BI^i}$: thus at least one parent has an $\mathbf{I^B}$ allele. Because he has a sister who is blood group A and a brother who is blood group O, this indicates that one parent has an $\mathbf{I^A}$ allele and both must have an $\mathbf{I^i}$ allele.

3. Expected progeny phenotypes will be 1 yellow : 2 cream : 1 white-coated.

$\mathbf{C^WC^Y} \times \mathbf{C^WC^Y} \rightarrow 1\ \mathbf{C^YC^Y} : 2\ \mathbf{C^WC^Y} : 1\ \mathbf{C^WC^W}$

4. Yes, 50% of F_2 plants will be expected to produce oval-shaped radishes.

[F_1 plants $\mathbf{S^LS^R}$. F_2 plants 25% $\mathbf{S^LS^L}$ (long), 50% $\mathbf{S^LS^R}$ (oval), 25% $\mathbf{S^RS^R}$ (round)]

5. No, because Charlie Chaplin could not have been the father. A child of blood group B has the genotype $\mathbf{I^BI^B}$ or $\mathbf{I^BI^i}$. No $\mathbf{I^B}$ allele could have been inherited from Joan Barry as she was blood group A ($\mathbf{I^AI^A}$ or $\mathbf{I^AI^i}$). Thus the $\mathbf{I^B}$ allele came from the child's father. Because Charlie Chaplin was blood group O his genotype was $\mathbf{I^iI^i}$.

6. Seven offspring would be expected to have normal legs.

creeper cockerel (**Cc**) × creeper hen (**Cc**) → 1 **CC** (lethal) : 2 **Cc** (creepers) : 1 **cc** (normal)

Thus surviving offspring are in the ratio of 2 creepers : 1 normal.

7. (a) Migratory behaviour seems to be determined by one gene with two incompletely dominant alleles.

(b) Let **M** = migratory allele; **N** = non-migratory allele

F_1 birds (**MN**) × non-migratory birds (**NN**) → 50% birds showing weak migratory behaviour (**MN**); 50% non-migratory birds (**NN**)

8. (a) All lentil seeds show M^1 marbled pattern.

(b) 50% seeds show M^2 marbled pattern, 25% dotted and 25% clear.

(c) 50% seeds show M^1 marbled pattern, 25% M^2 marbled pattern and 25% dotted.

9. Two platinum foxes to every one silver-coated fox.

platinum (**Pp**) × platinum (**Pp**) → 1 **PP** (lethal) : 2 **Pp** (platinum) : **pp** (silver)

10. It would appear the three coat colours are controlled by one gene with two incompletely dominant alleles, $\mathbf{C^R}$ and **C**

Parents	F_1 progeny	Parents' genotypes
(a) cremello × palomino	50% cremello; 50% palomino	$\mathbf{C^RC^R \times C^RC}$
(b) chestnut × palomino	50% chestnut; 50% palomino	$\mathbf{CC \times C^RC}$
(c) palomino × palomino	25% cremello; 50% palomino: 25% chestnut	$\mathbf{C^RC \times C^RC}$

11. (a) Recessive.

(b) If **D** = allele for normal hearing and **d** = allele for deafness: (i) **Dd**; (ii) **dd**; (iii) **Dd**; (iv) **dd**

12. (a) ¼ or 25%

(b) 100%

(c) $^{27}/_{64}$ (¾ × ¾ × ¾)

Let **C** = cloven-footed allele and **c** = mule-footed allele

All F_1 pigs are cloven-footed (**Cc**). ¾ F_2 pigs are cloven-footed (**CC** or **Cc**) and ¼ are mule-footed.

13. Let **B** = allele for black feathers and **W** = allele for white feathers.

erminette cockerel (**BW**) × erminette hen (**BW**) → chicken in the expected proportions of 1 black (**BB**) : 2 erminette (**BW**) : 1 white (**WW**).

(a) $^1/_{16}$ (½ × ½ × ½ × ½, because each egg has a probability of ½ of being erminette)

(b) $^1/_{256}$ (¼ × ¼ × ¼ × ¼, because each egg has a probability of ¼ of being white)

(c) $^1/_{256}$ (¼ × ¼ × ¼ × ¼, because each egg has a probability of ¼ of being white, and also of being black)

14. $^9/_{16}$ (¾ × ¾)

Chapter 4

1. Double heterozygote, or **PpHh** (if **P**, **p**, **H**, **h** represent the alleles for purple-stemmed, green-stemmed, hairy-stemmed and hairless stemmed respectively).

The approximate 9 : 3 : 3 : 1 proportions of the four phenotypes among the offspring indicates the heterozygous genotype.

2. $^3/_{16}$ of the progeny. Do a Punnett Square to show the possible fertilizations between gametes produced by two heterozygotes. Any individual of genotype **bbS-** will have brown hair and a short tail.

$Z^gZ^g \times Z^gW \rightarrow Z^gZ^g$ (white males), Z^gW (white females)

10. If male parent is solid colour and female parent barred, then all male chickens will be barred and all female chickens will be solid coloured.

    $Z^bZ^b \times Z^BW \rightarrow Z^BZ^b$ barred male chickens, Z^bW solid female chickens

Chapter 7

1. Linkage of the two loci is indicated by the larger number of parental phenotypes (brown and albino) compared to the recombinants (orange and black). If the two genes were on separate chromosomes, the cross would have produced approximately equal numbers of the four phenotypes (i.e. a 1 : 1 : 1 : 1 ratio).

2. Let **R** = resistance allele, **r** = sensitivity allele; **G** = green pod allele, **g** = orange pod allele.

   (a) F_1 phenotype is green pods and resistance to pea mosaic virus

   F_1 genotype is $\frac{\mathbf{R} \quad \mathbf{G}}{\mathbf{r} \quad \mathbf{g}}$

   (b) Possible phenotypes are green pods and resistance to virus; green pods and sensitivity to virus; orange pods and resistance to virus; orange pods and sensitivity to virus. The first and fourth phenotypes will be the most frequent.

3. (a) The unequal proportions among the four progeny classes, with greater numbers of the parental phenotypes (coloured fur, normal gait and waltzer albinos).

   (b) If **C** =allele for coloured fur, **c** = albino allele, **N** = normal gait, **n** = waltzing gait, the genotypes producing each of the four phenotypes are:

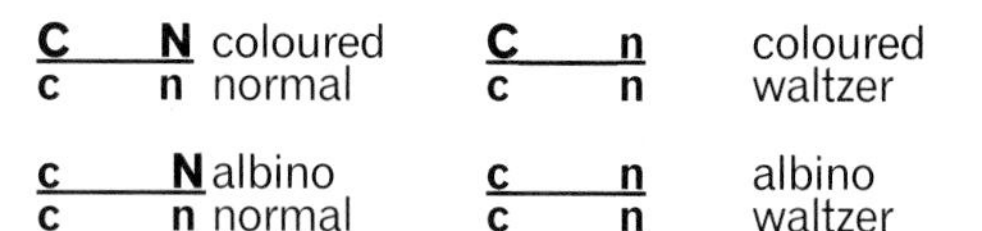

   (c) 50% albino waltzers and 50% albino normal gait mice

   $\frac{\mathbf{c} \quad \mathbf{n}}{\mathbf{c} \quad \mathbf{n}} \times \frac{\mathbf{c} \quad \mathbf{N}}{\mathbf{c} \quad \mathbf{n}} \rightarrow \frac{\mathbf{c} \quad \mathbf{n}}{\mathbf{c} \quad \mathbf{n}}$ and $\frac{\mathbf{c} \quad \mathbf{N}}{\mathbf{c} \quad \mathbf{n}}$

4. Equal numbers of the two parental phenotypes, i.e. the double recessive (produced by **aabb**) and the double dominant (produced by **AaBb**):

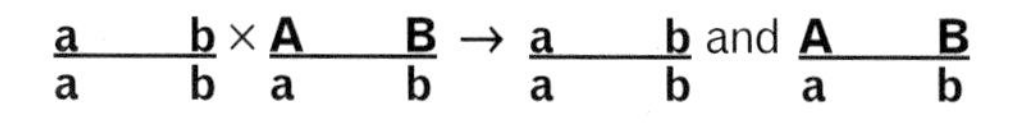

5. Independent assortment of two genes can occur if the two are far apart on the same chromosome. Thus, during prophase I of meiosis, a chiasma always forms between the two genes, producing recombinant gametes.

6. 7.7 cM, because 1% recombination indicates a map distance of 1 cM.

7. 26 cM

   $$\% \text{ recombinants} = \frac{\text{no. recombinants}}{\text{total number of progeny}} = \frac{21}{80} = 0.26$$

   (recombinant frogs are green, large-padded and blue, normal-padded)

8. The genes controlling colour of body and shape of antennae are on the same chromosome. Because only a few crosses yielded any recombinant progeny, flies with aristapedia antennae and grey bodies and normal antennae and ebony bodies, the two genes must be very close together.

9. (a) 860 **AaBb**, 140 **Aabb**, 140 **aaBb** and 860 **aabb** (because the two genes are 14 cM apart, 14% of the offspring are expected to be recombinants, i.e. to possess the genotypes **Aabb** and **aaBb**).

   (b) The same results as part (a).

10.

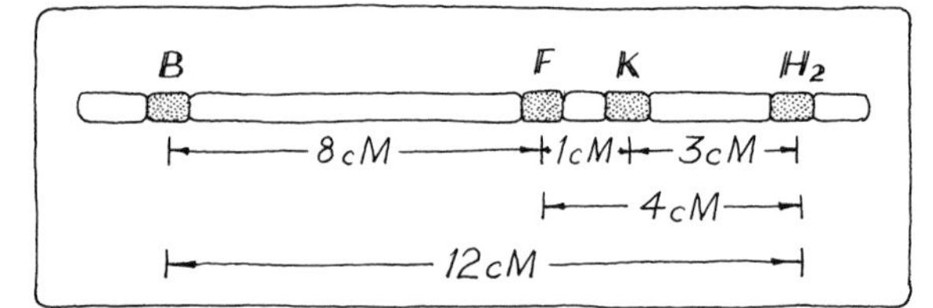

Chapter 8

1. (a) 45; (b) 69; (c) 47; (d) 48

2. (a) Inversion; (b) duplication; (c) duplication and inversion; (d) translocation

3. (a) Hybrid chromosome number = 29; allopolyploid chromosome number = 58

   (b) Hybrid will be infertile: there are 29 different chromosomes and so homologous chromosome pairing is impossible during prophase I of meiosis. Unbalanced gametes result. In the allopolyploid, produced by chromosome doubling in the hybrid, each chromosome will now be represented twice, pairing is once again possible during meiosis and balanced gametes will be produced, containing two copies of each chromosome. Thus fertile offspring can be produced.

4. Problems generally only result if the break point of an inversion or translocation occurs within a gene. However meiosis within the cells of an individual containing an inversion or a translocation can result in unbalanced gametes, and so unbalanced zygotes at fertilization. Thus problems can occur in subsequent generations.

5. $^1/_8$ gametes would be expected to contain all three chromosomes; $^3/_8$ gametes contain two chromosomes; $^3/_8$ gametes contain one chromosome and $^1/_8$ gametes contain no chromosome (considering the chromosome that is trisomic). This range of gametes results because of pairing problems at prophase I of meiosis. Three chromosomes cannot successfully pair.

6. Gene order is q s p r u t

7. One of the parents' cells (parent 2 below) contains a balanced reciprocal translocation between chromosomes 9 and 12. An unbalanced gamete from this parent was used at fertilization, resulting in the chromosome composition of the child shown in the diagram.

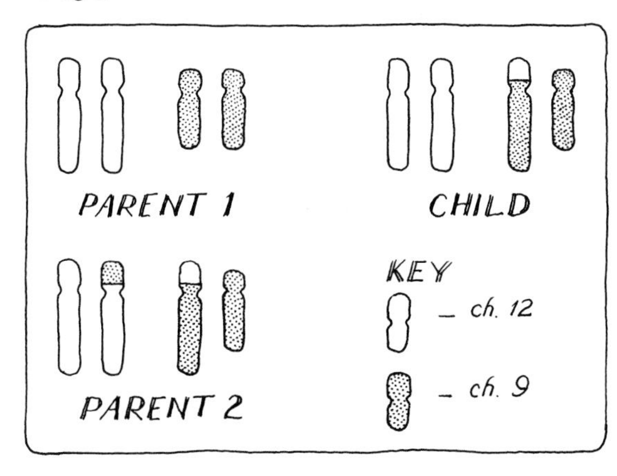

8. Twelve chromosomes per gamete. An autohexaploid has six chromosome sets in each cell, thus 12 chromosomes will be found in a haploid set.

9. An allotetraploid possesses four copies of each chromosome. Thus, if there are 44 chromosomes in each cell, there are 11 different chromosomes, or linkage groups.

10. Mother. A woman with Turner's syndrome is XO. If she is colour blind her genotype is $\mathbf{X^{cb}O}$. Thus she must have inherited her father's X chromosome.

11. 64.

Chapter 9

1. (a) Mean

(b) Distribution with the bigger variance

(c) 68.26% population lie within one standard deviation; 95.44% population lie within two standard deviations.

2.

Nest	Woodland weights (g)	$x-\bar{x}$	$(x-\bar{x})^2$	Scrubland weights (g)	$x-\bar{x}$	$(x-\bar{x})^2$
1	10.4	–0.11	0.012	9.01	–0.56	0.314
2	10.2	–0.31	0.096	9.86	0.29	0.084
3	10.6	0.11	0.012	9.71	0.14	0.02
4	9.4	–1.11	1.232	8.84	–0.73	0.533
5	11.0	0.49	0.24	9.33	–0.24	0.058
6	10.8	0.31	0.096	10.16	0.59	0.348
7	11.3	0.79	0.624	10.8	1.23	1.513
8	10.3	–0.21	0.044	8.94	–0.63	0.397
9	10.3	–0.21	0.044	9.66	0.09	0.008
10	10.8	0.31	0.096	9.43	–0.14	0.020

(a) Mean weight of (i) woodland nestlings is 10.51 g; (ii) scrubland nestlings is 9.57 g

(b) Variance of (i) woodland nestlings = 0.277; (ii) scrubland nestlings = 2.947

NB variance $= \dfrac{\Sigma(x-\bar{x})^2}{n-1}$

(c) Scrubland population has the larger standard deviation

NB standard deviation $=\sqrt{\text{variance}}$. Standard deviation of woodland population = 0.527; of scrubland population = 0.572

3. F_2

4. (a) $H^2 = 0.653$

$H^2 = V_G/V_P$; $V_P = V_E + V_G$. F_1 variance represents V_E; F_2 variance represents V_P.

Using F_1 and F_2 data: $4.96 = 1.72 + V_G$. $V_G = 4.96 - 1.72 = 3.24$. $H^2 = 3.24/4.96 = 0.653$

(b) Yes

5. (a) If the two varieties are subject to the same growth conditions, then individuals of the variety showing the low variance are likely to be genetically similar while individuals of the other variety have a range of different genotypes. Alternatively, both varieties could be genetically homogeneous, but the variety showing the higher variance could be subjected to a wider range of growth conditions.

(b) The variety showing the lower variance, as I could be more certain of the range of fruit weights I would obtain.

(c) Variety with the higher variance, as I am more likely to get some heavier fruit to include in my breeding programme.

6. (a) 2.28%. 16.4 min represents 2 standard deviations above the mean. 4.56% of children will complete the maze in a time 3 standard deviations or more above or below the mean (see Fig. 9.4). Thus, 2.28% of children will take more than 2 standard deviations.

(b) 84.13%. 50% of children will reach the centre in a time equal to or faster than the mean. 34.13% of children will reach the centre in a time within one standard deviation below the mean; 6.8 min represents 1 standard deviation below the mean.

(c) 15.87%. All children not falling within the range of answer (a).

7. Heritability would decrease. After many generations of selective breeding, the population would become genetically uniform, i.e. possess genes for large berries.

8. (a) All F_1 pumpkins of weight 4.5 kg.

(b) F_2 pumpkins of 3 kg, 3.5 kg, 4 kg, 4.5 kg, 5 kg, 5.5 kg and 6 kg in the ratio of 1 : 6 : 15 : 20 : 15 : 6 : 1.

9. (a) $H^2 = 0.69$. $V_G = 4.2 + 1.6 + 0.3 = 6.1$. $H^2 = V_G/V_P = 6.1/8.8$

(b) $h^2 = 0.48$. h^2 = additive genetic variance $(V_A)/V_P = 4.2/8.8$

10. (a) $h^2 = 0.5$ also equals selection response (R)/selection differential (S). $R = 15.5$ cm – 13 cm = 2.5 cm. $S = 18$ cm – 13 cm = 5 cm. Thus $h^2 = 2.5/5 = 0.5$

(b) A selective breeding programme is likely to be effective in increasing the length of the 8-week salmon

11. (a) $h^2 = 0.32$. $R/S = 8/25$. A moderate value for h^2 which indicates that a selective breeding programme will have some impact on increasing average body weight.

Chapter 10

1. (a) 0.8 (NB: frequency of dominant allele = p; frequency of recessive allele = q; $p + q = 1$. Because $q = 0.2$, and $p + q = 1$, $p = 0.8$)

(b) 0.32 [Use the Hardy–Weinberg law, which states that the three genotypes that can be produced by a monohybrid locus with two alleles are present in a population in the frequencies: p^2 (homozygous dominant), $2pq$ (heterozygote) and q^2 (homozygous recessive)].

2. 188 vipers. Non-poisonous vipers possess a homozygous recessive genotype, **vv**. The frequency of allele **v**, or q, is 0.25. Thus the frequency of the recessive genotype, or q^2, is $(0.25)^2 = 0.0625$. The population consists of 3000 vipers. Therefore, the number of non-poisonous vipers = 3000 × 0.0625 = 187.5, or 188 to nearest whole number.

3. Assume that the population is in genetic equilibrium, thus allele and genotype frequencies will be the same in the next generation.

(a) 81%

In order to calculate frequencies of individuals showing the heterozygous and homozygous dominant genotypes, values for the dominant and recessive alleles must first be calculated. 81% of individuals are homozygous recessive or q^2.

If $q^2 = 0.81$, $q = \sqrt{0.81} = 0.9$, thus $p = 0.1$

(b) 18%. Heterozygote frequency = $2pq = 2 \times 0.9 \times 0.1 = 0.18$

(c) 1%. Frequency of homozygous dominant = $p^2 = (0.1)^2 = 0.01$

4. (a) Phenotype frequency; (b) genotype frequency; (c) allele frequency

5. 1 in 139.

$q = 0.00362$; $p = 0.9964$. $2pq = 0.00721$, or 1 in 139 individuals.

6. (a) 0.43, or 43%

$q^2 = 0.1$. Thus, $q = \sqrt{0.1} = 0.316$ and $p = 0.684$. $2pq$ (heterozygote frequency) = 2 × 0.316 × 0.684 = 0.43

(b) 0.48, or 48%

$$\frac{\text{frequency heterozygote}}{\text{frequency individuals with dominant phenotype}}$$

$$= \frac{2pq}{p^2 + 2pq} = \frac{0.432}{0.468 + 0.432} = 0.48$$

7. 0.53

$0.78 = p^2 + 2pq$; thus $0.22 = q^2$. $q = \sqrt{0.22} = 0.47$

p (resistance allele frequency) = 0.53

8. (a) Genetic drift, migration, mutation, selection.

(b) Recessive alleles are also present in the heterozygote. Selection will either not act against these heterozygotes, or at a reduced level.

(c) Much lower; most individuals are homozygous – dominant or recessive.

(d) It is the only allele present at a given locus, which is therefore monomorphic.

(e) Because mutations generally occur at an extremely low frequency, in the order of 10^{-5} per generation. Thus it takes many generations for a substantial change in allele frequency to occur.

(f) Small population size.

9. (a) and (b) – see table below.

Phenotype	Number of individuals	Genotype	Genotype frequency	Number of alleles	
				B	G
Green	120	**GG**	0.6 ($^{120}/_{200}$)	240	
Brownish -green	60	**BG**	0.3 ($^{60}/_{200}$)	60	60
Brown	20	**BB**	0.1 ($^{20}/_{200}$)		40
Total	200			300	100

Frequency allele **G** $= \frac{300}{400} = 0.75$ Frequency allele **B** $= \frac{100}{400} = 0.25$

(c) Expected frequencies = p^2, $2pq$, q^2. Using the above values, the expected frequencies are:

GG = 0.562; **BG** = 0.375; **BB** = 0.0625

(d) Perform a χ^2-test:

Phenotype	Observed numbers (*O*)	Expected numbers (*E*)	*O* – *E*	$(O-E)^2$	$(O-E)^2/E$
Green	120	112.5	7.5	56.25	0.5
Brownish-green	60	75	–15	225	3.0
Brown	20	12.5	–7.5	56.25	4.5
Total	200	200			8.00

Null hypothesis: there is no difference between observed and expected numbers

Significance level = 0.05; degrees of freedom = 2; calculated χ^2 value = 8.00; critical χ^2 value = 5.99.

Because the calculated value is greater than the critical one, at a significance level of 0.05 and three degrees of freedom, the null hypothesis is rejected. The population of frogs is NOT in Hardy–Weinberg equilibrium when the colour locus is examined.

10. Use the same method as outlined in the answer to Question 9

Frequency of allele **A** = 0.56, allele **B** = 0.44

Expected genotype frequencies: **AA** = 0.31; **AB** = 0.49; **BB** = 0.19

To test for Hardy–Weinberg equilibrium, perform a χ^2-test.

Null hypothesis: there is no difference between observed and expected numbers

Significance level = 0.05; degrees of freedom = 2; calculated χ^2 value = 12.4; critical χ^2 value = 5.99.

Because the calculated value is greater than the critical one, at a significance level of 0.05 and three degrees of freedom, the null hypothesis is rejected.

The population of salamanders is NOT in Hardy–Weinberg equilibrium when the malate dehydrogenase locus is examined.

Chapter 11

1. (a) 400; (b) 200; (c) 45; (d) 400; (e) 155

If there are 200 nucleotide pairs in a fragment of DNA, there must be a total of 400 bases. If 45 of these are thymine, 45 will be adenine, because of complementary base pairing. Together these account for 90 of the 400 bases. The remaining 310 bases are cytosine and guanine, 155 of each.

2. (a) 3'-TGGCCATCTTAAGC-5'; (b) 5' to 3'

3. 23% are adenine.

If there are 27% guanine bases, there are also 27% cytosine bases. Together these account for 54% of the bases. 46% are adenine and thymine – 23% of each.

4. Opposite directions.

5. (a) 30.000 complete turns; (b) 300.000 nucleotide pairs

One complete turn of the DNA double helix occupies 34 Å (10 Å = 1 nm), and one nucleotide pair occupies 3.4 Å.

6. (b) and (c). Substitute some figures, e.g. 20% A, 20% T, 30% C and 30% G and work out the ratios.

7. (a) False; (b) false (complementary, not identical); (c) true (because of complementary base pairing; (d) false; (e) false (there will be 16% cytosine)

8. These terms refer to modes of DNA replication. Semi-conservative replication refers to the fact that each new DNA molecule possesses one conserved parental strand and one newly synthesized strand. Conservative replication refers to the suggestion in the 1950s that replication might produce one molecule with both parental strands and the other with both strands newly synthesized.

9. DNA polymerases can only synthesize in a 5' to 3' direction. Because of the antiparallel nature of the two DNA strands, if replication is proceeding in one overall direction, only one new strand can be synthesized in a 5' to 3' direction. The other strand is synthesized in segments in the apparently wrong direction, and the segments subsequently joined together.

10. 39% of the bases are also thymine. Thus 11% are cytosine and 11% are guanine. The DNA, therefore, contains an unusually high percentage of adenine and thymine.

11. (a) 344,560 nucleotides; (b) 58.57 μm (NB a base pair occupies 3.4 Å)

Chapter 12

1.

DNA	RNA
Pentose sugar: deoxyribose	Pentose sugar: ribose
Pyrimidines: cytosine and thymine	Pyrimidines: cytosine and uracil
Double-stranded	Single-stranded
One molecule contains millions of nucleotides	One molecule contains hundreds or thousands of nucleotides
One function: contains encoded information	A variety of functions: mRNA carries encoded information of one gene; tRNA carries amino acids; rRNA constituent of ribosomes
Located in nucleus and two organelles, the mitochondria and chloroplasts	Made in nucleus, but then found in cytoplasm: all three types also found in mitochondria and chloroplasts

2. (a) tRNA; (b) mRNA, rRNA and tRNA; (c) tRNA; (d) mRNA; (e)mRNA; (f) tRNA

3. (a) 3′-CCUUGGGUC-5′; (b) CAG

4. Deletion of a base changes the reading frame. All codons after the mutation point are altered; thus all encoded amino acids are different and so the resulting protein is rendered non-functional.

5. Transcription: alignment of ribonucleotides against the template DNA strand; translation: tRNA anticodon binding to mRNA codon.

6.

mRNA	tRNA
Single strand of RNA	Formed as single strand; folds into cloverleaf shape with double-stranded regions
Main bases A U C G	Contains additional bases to A U C G
Contains the encoded genetic information	Binds to, and transports amino acids from, cytoplasm to ribosomes

7. Nucleotides, codons, exons, genes, chromosomes, genomes.

8. 579 nucleotides (three nucleotides encode one amino acid)

9. 1 C; 2 J; 3 G; 4 I; 5 H; 6 A; 7 E; 8 D; 9 F; 10 B

10.

<table>
<tr><td>T</td><td>C</td><td>G</td><td>A</td><td>C</td><td>C</td><td>T</td><td>G</td><td>A</td><td>C</td><td>T</td><td>T</td><td rowspan="2">DNA</td></tr>
<tr><td>A</td><td>G</td><td>C</td><td>T</td><td>G</td><td>G</td><td>A</td><td>C</td><td>T</td><td>G</td><td>A</td><td>A</td></tr>
<tr><td>A</td><td>G</td><td>C</td><td>U</td><td>G</td><td>G</td><td>A</td><td>C</td><td>U</td><td>G</td><td>A</td><td>A</td><td>mRNA transcribed</td></tr>
<tr><td>U</td><td>C</td><td>G</td><td>A</td><td>C</td><td>C</td><td>U</td><td>G</td><td>A</td><td>C</td><td>U</td><td>U</td><td>tRNA anticodon</td></tr>
<tr><td colspan="3">serine</td><td colspan="3">tryptophan</td><td colspan="3">threonine</td><td colspan="3">glutamine</td><td>encoded amino acid</td></tr>
</table>

$(X^2{}_{critical})$

Probability Level of Significance

degrees of freedom (df)	0.1	0.05	0.025	0.01
1	2.7	3.8	5.0	6.6
2	4.6	5.9	7.9	9.2
3	6.2	7.8	9.3	11.3
4	7.7	9.4	11.1	13.2
5	9.2	11.0	12.8	15.8
6	10.6	12.5	14.4	16.8
7	12.0	14.0	16.0	18.4
8	13.3	15.5	17.5	20.0
9	14.6	16.9	19.0	21.6
10	15.9	18.3	20.4	23.2
11	17.2	19.6	21.9	24.7
12	18.5	21.0	23.3	26.2
13	19.8	22.3	24.7	27.6
14	21.0	23.6	26.1	29.1
15	22.3	24.9	27.4	30.5
16	23.5	26.2	28.8	32.0
17	24.7	27.5	30.1	33.4
18	25.9	28.8	31.5	34.8
19	27.2	30.1	32.8	36.1
20	28.4	31.4	34.1	37.5
21	29.6	32.6	35.4	38.9
22	30.8	33.9	36.7	40.2
23	32.0	35.1	38.0	41.6
24	33.1	36.4	39.3	42.9
25	34.3	37.6	40.6	44.3
26	35.5	38.8	41.9	45.6
27	36.7	40.1	43.1	46.9
28	37.9	41.3	44.4	48.2
29	39.0	42.5	45.7	49.5
30	40.2	43.7	46.9	50.8
31	41.4	44.9	48.2	52.1
32	42.5	46.1	49.4	53.4
33	43.7	47.4	50.7	54.7
34	44.9	48.6	51.9	56.0
35	46.0	49.8	53.2	57.3

Index